Little Britches

Father and I Were Ranchers

**Center Point
Large Print**

**This Large Print Book carries the
Seal of Approval of N.A.V.H.**

Little Britches

Father and I Were Ranchers

Ralph Moody

Center Point Publishing
Thorndike, Maine

TO MY FRIEND
A. MARSHALL HARBINSON
without whose encouragement and instruction this story
would not have been written

This Center Point Large Print edition
is published in the year 2004 by arrangement with
University of Nebraska Press.

The text of this Large Print edition is unabridged. In other
aspects, this book may vary from the original edition. Printed in
Thailand. Set in 16-point Times New Roman type by
Bill Coskrey and Gary Socquet.

ISBN 1-58547-408-8

Library of Congress Cataloging-in-Publication Data

Moody, Ralph, 1898-
    Little britches : father and I were ranchers / Ralph Moody.--Center Point large print ed.
    p. cm.
Originally published: Lincoln : University of Nebraska Press, c1950.
    ISBN 1-58547-408-8 (lib. bdg. : alk. paper)
    1. Moody, Ralph, 1898- 2. Moody, Ralph, 1898---Family. 3. Moody family.
    4. Ranchers--Colorado--Biography. 5. Ranch life--Colorado--History--20th century.
    6. Fathers--Colorado--Biography. 7. Colorado--Biography. I. Title.

CT275.M5853A3 2004
978.8'031'092--dc22

                                                                    2003019993

CONTENTS

1

We Move to Colorado

I NEVER REALLY KNEW FATHER VERY WELL TILL WE moved to the ranch on the Fort Logan—Morrison road, not far from Denver. That was just after my eighth birthday—right at the end of 1906. When we lived in East Rochester, New Hampshire, he worked in the woolen mill, but it wasn't good for his lungs. He was sick in bed the winter before we moved—the one after Hal was a year old.

Cousin Phil lived in Denver, and came to see us the next spring, right after Father got well enough to go back to work. I liked him a lot. He had a gold front tooth, and wore a derby hat cocked way over on his right ear. And he sold gold-mine stock.

One afternoon when Grace and I got home from school, he and Mother were talking in the parlor. I didn't have much chance to listen, because Mother told Grace and me to take Philip and Muriel outside to play till suppertime. But I did hear Cousin Phil say, "Why, Mame, there just isn't any work at all to ranching in Colorado. We have three hundred and sixty-five sunshiny days in a year, and all a man has to do is toss out seed in the spring and harvest his crop in the fall. With my connections, I could make a deal to put you folks on one of the finest ranches in the country, where you'd have all the milk, butter, and eggs you could eat, and half of all the crops you could raise. Why, in one year Charlie'd be a new man—and make as much money as he'd make here in East Rochester in a lifetime."

I guess Father and Mother believed what he said, because there were letters from him all through the summer and fall. Then, just after Christmas, we had our auction and took the train for Denver—all seven of us: Father and Mother and Grace, Muriel, Philip, Hal, and I. Grace was older than I was, but the rest were younger. All the way out on the train, I kept guessing how big the house and barns on our ranch would be, and how many hundred horses and cows there'd be on it.

It was late when we got to Denver, so we rented a room in a little hotel on Seventeenth Street. The next day, Cousin Phil lent us his rubber-tired buggy and Prince, his sleek little seal-brown driving horse. Father let me go to see our ranch with him and Mother. I didn't really have to ask him to let me go. I guess he just knew how much I wanted to and said to Mother, "Do you think there'd be enough room for you and the baby if we squeezed Ralph in between us?"

We could see our new house from a couple of miles away. We knew it must be ours, because Cousin Phil had told us it was three and a half miles west of Fort Logan—the first house on the Morrison wagon road. From the hill beyond the Fort, it looked like a little doll-house sitting on the edge of a great big table, with a brown tablecloth smoothed out flat all around it. It was right near the edge of the mesa, where the land started dipping northward into Bear Creek Valley. Away toward the south there were brown, rolling hills, as though the tablecloth had been wrinkled a little. And not far beyond it, toward the west, the hogbacks rose like big loaves of golden-brown bread sitting on the table. High above them the snowcaps of the Rockies glistened in the afternoon sunshine.

As we came nearer, it looked less like a dollhouse and more like just what it was: a little three-room cottage that had been hauled out from Denver. It was propped up on four cribs of mover's timbers, and sat at the corner of an unfenced quarter section of barren prairie land. The chimney was broken off at the roof and most of the windows were smashed. When we turned off the wagon road, a jack rabbit leaped out from under the house and raced away through the clumps of cactus and soapweed. But it was going to be *our* ranch—it looked all right to me.

Father and Mother didn't say a word, but when I looked up, the bunches of muscle at the sides of Father's jaws were working out and in. They always did that when he was trying not to get mad. Mother's face was as white as Hal's stocking cap, and her eyes looked as though she were going to cry, but she didn't. After Father tied Prince and helped Mother out of the buggy, he held me up so I could look in one of the windows. There wasn't much to see, except that the floor was covered with broken glass, and plaster that had fallen off the walls and ceiling.

While I was still looking in the window, Mother said, "Charlie, I don't see how in the world we can do it . . . with only three hundred and eighty-seven dollars. I thought, of course, there'd be good buildings and stock and machinery on it. We've got a lot of planning to do."

Her voice sounded hoarse, and seemed to be coming from way down in her throat.

Father didn't say anything until he had stood me down and taken Hal from Mother. Then he put his arm around her shoulder and hugged her up against him, Father was real tall, but slim, and Mother's head fitted in under his chin. "There's only one thing to plan about, Mame," he said, "and that's getting tickets home while we've still got the money. I won't have you live in any such God-forsaken place as this."

They stood that way for two or three minutes while Father's hand patted up and down on Mother's shoulder. And there wasn't a sound, except that dry little cough that Father had then. When Mother lifted her head, her lips were pressed tightly together, and her voice wasn't trembly any more. "The Bible says, 'Trust in the Lord and do good; so shalt thou dwell in the land, and verily thou shalt be fed.' The hand of God has led us here; we have set our shoulders to the wheel, and we will not turn back."

The next two days were not good ones for me. We stayed right in the hotel room, and Father was away from early in the morning till long after dark. There was nothing for us children to do, and I guess we made Mother nervous. Grace and I had two or three squabbles. She was the oldest, nearly two years older and always smarter than I, and I always got the spankings. When Father came in, the second night, Mother said, "Ralph can neither stay out of mischief himself, nor let any of the others. I declare, I shall go frantic if I have to have him cooped up in this little room another day."

Father didn't scold me at all, though. He just put his arm around Mother's shoulder, and said, "There, there, Mame. I know how hard it is for all of you. We'll get out of here just as soon as we can."

The next morning I went with Father, and we got a team of horses, and a wagon and harness. They were all kind of old and secondhand, but they were *ours* and I was proud of them. Father let me name the horses. I called the white one Bill, and the other one Nig.

We got up before daylight every morning for the next two weeks, Sunday and all. First, we'd pick up any of the bargains Mother had found for the house, then buy secondhand lumber, plaster, glass, and other things we needed, on our way out to the ranch. And Father would never stop working till it was so dark he couldn't see to drive a nail.

He got a man to come and help him dig the well, and some days Cousin Phil drove out and worked on the barn with us. It just had three sides and a corrugated iron roof. By Thursday night the barn was all finished, and Father had built a new chimney, patched the places where the plaster had fallen off, put glass in all the windows, and made front and back steps for the house. My part of the job was to sweep up all the broken glass and plaster, and pile up all the little pieces of board beside the back steps. There was nothing left to build but the privy.

It was just five o'clock when Father and I drove up to the depot platform Friday morning. After the baggage man had found our two trunks, we went around to the hotel and picked up everything we wouldn't need for one more night in the room. We had to tiptoe in to get the things, because all the other children were still asleep in the shakedown. I hadn't seen any of them when they were awake since the Sunday before.

The sun looked only about a foot high when we stopped at a feed barn on the outskirts of Denver. Father bought a sack of oats and four bales of hay there, and we fed the horses. While they were eating, we went to a little store up the street, and Father bought a pail of milk and a whole custard pie. We ate it sitting on the curb beside our wagon. Father knew just how to buy a good breakfast.

At Fort Logan we stopped at Mr. Green's general store, and bought more groceries than I thought we could ever eat. There was a barrel of flour and a hundred-pound sack of navy beans; and salt pork, molasses, sugar, rice, and a whole case of evaporated milk. We got out to the ranch long before noon.

Cousin Phil drove out that afternoon and helped us with the privy. Father sawed the two-by-fours and spiked them together. Then, while he was making the door and the seat, Cousin Phil cut boards, and nailed them on the sides and roof. He cut some of them too long, and some too short. After a while he tossed the folding ruler over to Father, and said, "This confounded rule isn't accurate, Charlie."

Father folded the ruler up and put it in his pocket. He didn't say anything till Cousin Phil had gone for another board, then he said to me, "If you just remember to measure twice and saw once, you'll get along all right."

The last thing we did was to tie Bill and Nig in their new barn, and Father hung the harness on spikes he had driven high up on the barn wall. He said that was so the coyotes couldn't gnaw on them during the night.

The sun had gone down, and the whole sky beyond the mountains looked as though it were on fire. I looked back at our ranch as Cousin Phil drove us in to Denver, and I wouldn't have traded it for anything else on earth.

A narrow-gauge spur of the Colorado and Southern left the main line at Petersburg junction, then followed Bear Creek to Fort Logan, where it climbed to the tableland, and ran west to Morrison. It was a single-track line, and crossed the center of our ranch. There were no passenger trains, but one freight each day had a passenger car hooked on the end of it.

We were all down at the depot early Saturday morning, and Father asked the conductor if they would stop the train so we could get off on our own place. As we were climbing down, the engineer blew three sharp toots on his whistle, and we all looked toward where he was pointing at something on the track ahead.

A quarter of a mile west there was a deep gulch where, after storms, water from the hills had cut the land away in running off to the creek. The railway crossed it on a high trestle, and something that looked to me like a big black whale was floundering around in the middle of it.

Father ran toward the front of the train, and I ran after him. The engineer cupped his hands around his mouth, and shouted, "Horse through the trestle. Is it yours?"

Father motioned for Mother to go to the house as he swung up on the engine. But he could run so much faster than I that the train started moving when I was at the back

13

end of the first box car. After going everywhere Father had for the last couple of weeks, I didn't want to be left behind then, so I grabbed hold of the foot rods and pulled myself up. I heard Mother scream, "Ralph!" but I held on tight, and the train didn't stop till it got to the trestle.

Nig's four legs were down between the crossties, and he was thrashing around like crazy. The first thing I wanted to do when I saw him was to run for home, but I couldn't pull my eyes away. His hind end was toward us and there were big bloody patches on his thighs. He was out nearly twenty feet on the trestle. Between him and us there was more blood on the crossties, and big clumps of white hair. I knew Bill had been in there, too, but he was nowhere in sight. I peeked down into the gulch and there he was—stretched out on a little patch of snow near the bottom. The snow was half covered with dust and tumbleweeds. And big blotches of bright red blood showed between them.

Nig would thrash and jump until he was all tired out, trying to pull his legs up from between the ties. Then he'd fall back and pound his head against the track. I was sure he would kill himself any minute. All the men from the train came running up to the trestle, but Father was the only one who seemed to know what to do.

There was a sign on a four-by-four post beside the track, right near the end of the bridge. Father wrenched it out of the ground, smashed the sign off, and ran out on the trestle toward Nig. I didn't want to see Father kill him, so I covered my eyes with my hands. There was a hollow thud, like a wooden tub hit with a stick. When I dropped my hands, Nig was lying perfectly still.

Father called to the men standing at the end of the trestle. His voice wasn't quiet then, as it usually was, but he didn't yell. It was big and deep like the ring of a church bell. "Bring chains and anything you can find to pry with," he called. "We'll have to hurry a bit; he'll come around in two or three minutes."

All the men started running around like ants when you plow into their nests. In another minute they went swarming out onto the bridge. They were all chattering like magpies, and some were yelling. Father's voice rang through the hubbub, deep and strong: "Run that chain under here! Take a pry across the top of that rail! Here, big fellow, heave up on his head. . . . Wait for the word!" He sounded the way I had always imagined George Washington must have, and I was proud he was my father. I saw him crouch with his back against Nig's hind end. He pulled the long tail over his shoulder and cried, "Up!"

Nig's legs came up through the ties with a rush. He must have come to at the very second they got him up. He thrashed and the men jumped away from him. In another second he toppled over the side of the trestle. There was a dull thud when he landed. It was then I realized that my pants were wet.

Father vaulted over the side of the trestle near our end and disappeared into the gulch. I didn't dare to look down. In a minute or two his voice came up. "No broken legs, and they're breathing well. I think they'll make out all right. Thanks!"

The trainmen didn't seem to care about anything except having the track clear. The engineer climbed back on the engine and tooted the whistle. In another minute the train had gone, and I was left all alone. Father came up over the edge of the gulch and picked me up. He didn't mention my pants, but unbuttoned his reefer and wrapped me inside it. "I'm sorry you had to see it, Son," he said, "but it's that sort of thing that makes a fellow into a man. We'll go get some bandages and see what can be done for them."

Carrying me to the house, he said, "Sometimes these things seem awful hard to take, but maybe they all happen for the best. Now you children will know that bridge is dangerous. It might have been one of you that fell off it." After that he pointed out a jack rabbit that was scurrying away along the track, and to a single, stunted cottonwood tree near the far end of our land. "There," he said, "who says we haven't got a wood lot on our place? Perhaps, with enough irrigation water, it will grow into a fine big tree. It would have been nice if they'd put the house by the tree, wouldn't it?"

When Father brought me into the house, Mother had a fire going in the cookstove, and everybody was standing by it getting warm. She looked up at Father, and her underlip was trembling. "Are they both dead, Charlie?" she asked.

"No," Father said, "they're both living. I don't know how badly they're hurt, but there don't seem to be any bones broken." He didn't unwrap his coat from around me, but whispered to Mother. Then we went into the front room where the trunks were, and she closed the door. They were the only ones who ever found out about my pants. And it never happened again.

While Mother ripped an old sheet into bandages, Father went out to look around the barn. When he came back, he said, "Coyotes. Must have closed in and frightened them about daylight. There's plenty of sign. What have you got for an antiseptic?"

Mother put her hand up to her mouth. "I don't think there's a thing here, except a couple of bichloride tablets."

"Never mind," Father said. "Ralph, you bring the bandages. I've got a can of axle grease in the wagon."

I hadn't expected him to take me with him after my accident, and pulled my coat on as fast as I could. I was afraid both horses might die before we could get back, and wanted Father to run, but he wouldn't. You could ask him all the questions you wanted to; he never got cross. So I said, "Why didn't it kill Nig to pound his head on the track? Do you think Bill pounded his head, too? Father, how did Bill get out when Nig couldn't?"

"Well," Father said, "Nig hadn't been in there long enough to do himself much damage. The blood was all fresh and bright, so they must have fallen in less than fifteen minutes before we got there. Nig pounded his head because he was frantic. Bill had no reason to do it, because he could get out. From the marks on the track, I'm sure that one of his hind legs didn't go through at all, and that he braced himself with his head to pull his front legs out. I'll show you when we get there."

Father whistled when we got near the edge of the gulch. He was so much taller than I that he could see down into it sooner. I ran to the edge. Bill and Nig were cropping grass around a wet spot. Nig was limping, but

Bill didn't seem to mind the blood that was oozing from torn places on his thigh and forelegs.

As soon as Father saw that the horses were up on their feet, we went over to the trestle. He picked me up and, after looking up and down the track, walked out on the bridge. Then he scrootched down and showed me all the marks on the crossties. "Almost everything that happens leaves its telltale marks," he said. "If you teach yourself to see all the marks, you can always read the story." Then he had me wait while he went down into the gulch and led the horses out. He said that since they were on their feet, we could do a better job of dressing them at the house.

Mother came out to help with the horses when we got back. She was always good when there was sickness. She took scissors and started clipping hair from around the torn places on Nig's forelegs. "I'm worried about this one, Charlie," she said. "He must be badly hurt to limp so."

Father was poking his fist up against Bill's belly. "I'm not worried much about him," he said, "but I'm afraid this one may be done for. I don't like the way he's drawn up in the loin."

2

NEIGHBORS

FATHER HAD JUST LED THE HORSES BACK TO THE BARN when a man drove into our yard with a pair of fast-stepping bays. He drove right past the house and swung around in a circle. His horses didn't slow down all the way around, but pulled up beside us with the pole of the buckboard pushing their collars way up high on their necks. The man wasn't quite so old as Father, but he was as tall, and a lot heavier. He stepped out of the rig without putting the lines down, and held his right hand out to Father. "I'm Fred Aultland, your next-door neighbor, a mile up the line," he said.

After Father shook hands and told him he was Charles Moody, Mr. Aultland held his hand out to me. I tried to take hold of it as Father did, but it was too big, and I only got hold of three fingers. "And mine's Ralph Moody," I said; "I like you." I did like Mr. Aultland right from the start.

"And I like a man that speaks his mind," he told me. Then he said to Father, "Hear you had a little hard luck with your team, and thought I'd drop in to see if I couldn't lend a hand. I got half a dozen teams standing around eating their heads off at this time of year. Better let me lend you one till yours gets back on its feet again."

Father said, "Thank you, Mr. Aultland, but I believe we'll make out all right. Of course, these fellows will be stiff for a few days, but I haven't got much hauling to do, and I don't think the black is hurt very bad."

Mr. Aultland said, "Hell, Charlie, don't call me Mister; my name's Fred." He stepped over near Bill and pushed his thumb down hard on his back along toward the hip. He looked at Father and lifted one eyebrow. All he said was, "Kidney?"

He and Father talked about horses and kidneys for a while. Then Mr. Aultland said he was going to Fort

Logan, and asked if there was anything he could bring us. Father sent me to ask Mother, and she told me the name of some kind of salve for the horses, but when I got back outside I had forgotten it. Mr. Aultland was already on his buckboard. He said, "Never mind, I'll get you some stuff that works wonders with galls and wire cuts." Then he let the lines go loose just half a second, and his team was away like Santa Claus's reindeers.

When Mr. Aultland came back, his horses were still running as fast as they were when he left. He drove around the circle, as he had before, and pulled up right beside the back steps. Father had gone to see if he could find the stakes that marked the corners of our land, so Mother went to the door. Mr. Aultland gave her a quart jar of blue-colored salve, a big square package, and a *Denver Post*. He said, "Tell Charlie to lay this stuff on over those sores good and heavy. It's got blue vitriol in it, but tell him not to be afraid of it. It'll dry those sores up quicker than anything else."

Mother looked at the package, and Mr. Aultland grinned. "Just a little baker's bread for the kids. I figured you'd have both hands full without bakin' for a couple of days."

Mother thanked him, and asked how much we owed him. "Forget it," he said. "Bessie or Mother'll prob'ly be down to borrow something off of you before the week's out."

"Thank you ever so much, and tell them I shall be delighted to see them," Mother said, as she turned away.

But Mr. Aultland called, "Say, I don't see any cows

around here. What you going to give these kids for milk?"

We were all watching out through the window. Mother's face got red as could be, and she said, "Oh, we have a whole case of evaporated milk; they'll be all right."

"That stuff's only good for chuck wagons," he said. Then he yelled, "Hey, Ralph. Get your jacket on and take a ride with me."

I just got a glimpse of the headline in the newspaper as I was getting my coat on. It said, "Man Killed by Mountain Lion at Moffat." Then Mother put it up on the lamp shelf.

When I climbed on the buckboard beside Mr. Aultland, he reached over and slapped me on the leg. It was a good hard slap, but I liked it. As we tore out of our yard, he asked me if I'd ever driven a team. I told him, yes, Father let me hold the lines when we were bringing out the lumber. He passed his reins right over to me, and said, "Here take ahold of 'em. Better wrap 'em around your hands once; you ain't very stout yet."

He showed me how to wrap the lines around my hands so they wouldn't slip, and told me to hold them up tight. The long-legged bays were running like sixty, and I was scared. I pulled on the lines as hard as I could, but all that happened was that my bottom slipped forward on the seat. Mr. Aultland put his arm around me and held me back so I could pull harder. He said, "Betcha my life you'll make a horseman. If you was my kid, I'd put a box in front here so you'd have something to brace your feet against."

As we got close to his house, he gathered both my hands inside one of his and helped me pull. The bays only slowed up a little, and the hind wheels of the buckboard slewed way around when we turned into his driveway. Aultland's house was four times the size of ours, and there was a big red barn, and corrals, and the fields beyond were knee-deep with brown stubble.

A tall, pretty girl came out to meet us when we stopped by the back door of the house. She had reddish-brown hair, and her eyes were the same color as a brand-new penny. She must have been nineteen or twenty. "Sis," Mr. Aultland said, "this is our new neighbor. There's a whole parcel of kids and they haven't got a cow. How about taking them over some milk? The woman seems to be right nice, and said she'd be glad to see you."

While the girl was asking me what my name was, and telling me hers, Mr. Aultland tied the horses to a hitch rack, and went off to the barn. She said her mother was frying a batch of doughnuts, and asked me if I wouldn't like to come in and have a hot one. I said, "Yes, I would, Miss Aultland. We haven't had any hot doughnuts since we left East Rochester."

She laughed and said, "Don't you dare call me Miss Aultland—that makes me sound like a schoolmarm. You call me Bessie. Come on now, we'll get some doughnuts."

Mrs. Aultland was as nice as Bessie. She wasn't very tall, but fat, with wavy gray hair. When I told her I liked Bessie and her husband fine, she laughed, and tweaked my ear. "That's the finest compliment I've had in years," she said, "but don't you let Fred fool you. He's just my

22

little boy, only he's big. He ain't even thirty yet. And don't you go calling him Mr. Aultland; it'll get him stuck-up. You call him Fred."

Bessie didn't let me drive going back. Maybe she didn't know I wanted to. She and Mother got along fine. I went out to the barn where Father was putting some of the blue salve on Bill and Nig. When we came back to the house, Bessie was saying to Mother, "I'm not going to keep saying Mrs. Moody. What shall I call you?"

Mother laughed and said, "That's just the way I'd like to have it. My name is Mary, but nobody ever calls me that. When I was a girl, they used to call me Molly."

Bessie said, "All right. Molly it is. I'll be seeing you often, Molly." She picked up the reins and was gone.

While we were eating supper that night, the coyotes began to howl. It sounded as though there were dozens of them; some close by, and some far away. It made shivers run up and down my back, and I think it did the same thing to Mother. As soon as supper was over, Father got up and took the lantern from the nail by the door. As he turned up the globe, Mother put both hands up to her cheeks and said, "Charlie, you're not going out there! I won't let you go out there!"

Father had lit the lantern. He set it down, and took Mother in his arms. "Mame," he said, "we'll have to face the situations we find in this country. These fellows can't be too dangerous, or Aultland would have warned us. If the horses were in shape to defend themselves, I wouldn't go. But they're not. I've rolled the wagon across the open side of the barn, so they can't break out again. Coyotes are said to be afraid of a light. I've got to hang this lantern on the wagon."

He picked up the lantern and went out. Mother stood in the open doorway, and we watched the lantern till it disappeared around the barn. The coyotes' howling stopped. In a few minutes Father was back, and said everything looked all right at the barn. Then the howling started again. Mother was still fidgety, and asked, "Where is Moffat, Charlie?"

Father looked at her and answered, "Moffat? Oh, it's in the mountains west of here somewhere. Why?"

"Oh, nothing," Mother said. "I just wondered, that's all."

Mother put the smaller children to bed while Grace and I did the dishes, and she made us go just as soon as we were finished. The coyotes stopped howling after a little while, but we couldn't go to sleep. Grace whispered over and asked me if I thought the mountain lions had come down and frightened the coyotes away. I was afraid they had, but I told her not to worry, because Father wouldn't let them get us. He must have heard us whispering, because he came to the door and said he didn't want to hear any more whispering. Father always meant what he said, so we kept quiet, and I guess we went to sleep pretty soon.

The moon was way over toward the mountains when something woke me up. It woke everybody else, too. My heart was pounding so hard I thought it was going to jump out. Then there was a clatter at the back of the house. Something had knocked the pile of firewood over. Grace shrieked, "The mountain lion!" and all the younger children yelled as though the lion had them by the ears.

Father leaped out of bed and ran to the kitchen for the lantern. I guess he thought Grace had really seen a lion. Mother rushed from one window to the other, slamming them down tight and crying, "Don't go out, Charlie! Don't go out! He'll kill you the way he did the man at Moffat."

Father didn't go out. He sat on the edge of the girls' bed, with Muriel on his knee and one arm around Mother, while he told us there wasn't a bit of danger. Everybody stopped crying pretty soon, but they were all holding their breath as I was. It was so still it almost hurt. But only for a few minutes. Then the most terrible noise I had ever heard came from right outside our window. We were all too scared to make a sound, till I heard Mother whisper, "Oh, God," and knew she was praying. A few minutes later there was an awful racket at the barn. We heard one of the horses squeal, and the sound of heels thudding against boards. Mother had to read to us a long time before we went back to sleep.

At daylight Father went to the barn to see if we had any horses left. In a couple of minutes he was back at the kitchen door, laughing and calling us all to come out to the barn. Cold as it was, Mother let us go without caps or coats. Standing between the two big horses was a Rocky Mountain canary—a little donkey, not much taller than I. He had been our mountain lion of the night, and had squeezed into the barn past the tail gate of the wagon.

That Sunday afternoon our new neighbors came to call. None of them came into the house, but they sat in their buggies and talked a little while. First were the Corcorans. They lived on the same road we did, up west of Fred Aultland's place. Mrs. Corcoran was a little bit of a woman, and did most of the talking—she had a high, sharp voice. Mr. Corcoran was a kind of round-shouldered man with a beard. She didn't let him talk very much. She started asking all kinds of questions about where we came from, and whether we'd bought the ranch, and was Father a lunger, and did we want to buy a cow.

Mother pulled her lips right up tight, but Father began telling them about the donkey—as if he hadn't heard a single question. I don't think Mrs. Corcoran liked it, because she said, "Robert, them cows of mine needs a load of hay before milkin'. It's time you was gettin' at it." Mr. Corcoran didn't say anything. He just fished on the reins a little, and the horses started moving. As they drove out of our yard, Mrs. Corcoran called back, "I hope you folks make out better than them Yankees that moved onto the Peterson place."

The next ones who came were the Aldivotes. They lived down near Bear Creek, behind Corcoran's place. They were nice people. Quiet. And didn't seem to find it easy to talk. They'd heard about Bill and Nig falling through the trestle, and I guess they just came to tell us they were sorry.

It was nearly sunset before Carl Henry drove in with Miss Wheeler. The schoolhouse was in the far corner of our section, and Carl's house was in the section beyond. He was an old bach—he must have been nearly thirty—and Miss Wheeler was the schoolteacher. She was prettier than Bessie Aultland.

At first they talked about Grace and me going to school. Then about horses, and fences and ranching. After a while, Mother told them how scared we'd been the night before, and asked Carl how much danger there was from coyotes and mountain lions. He laughed, and told her that the donkey was just about as dangerous as the coyotes, and that he had never heard of a lion coming that far down from the mountains. When they left, they took me as far as Aultland's to get the milk.

I hadn't much more than started back when I heard horses running behind me. I looked around, and there were four honest-to-goodness cowboys coming down the road. They wore ten-gallon hats and leather chaps with bright silver disks on them. As they came closer, I could see holsters with six-shooters in them, strapped to their waists. I was so busy watching that I forgot to move.

They didn't slow up a bit till they were right beside me, then they skidded their horses to a stop on the hard adobe road. One of them leaned over and said, "Want a lift, Sonny?"

I almost bit my tongue before I could make it say, "Sure I do."

He leaned so far out of his saddle that he took the milk bucket right out of my hand without my lifting it. Then he passed it to one of the other fellows, and swung me up behind his saddle by one arm. I had hardly landed when the horses started off. My cowboy said, over his shoulder, "Hang on if you want to burn some trail." I dug my fingers in under his cartridge belt. Somebody yelled,

"Yipeee!" and we were off like scared rabbits.

Mother used to recite "The Charge of the Light Brigade." With all the guns and running horses, I was sure I was in it. They put me down right at our back steps and raced away. There wasn't a drop of milk spilled when the cowboy passed me the bucket.

3

FIGHT, MOLLY!

MONDAY MORNING GRACE AND I WENT TO SCHOOL, and the attendance went up a fifth. Bessie Aultland came for us and drove us over to the little brick schoolhouse, a mile and a half from home.

When Bessie took us in to Miss Wheeler, she said, "I tried to tell Molly just to let them come in overalls and frock, not to get them all dressed up like they were going to Sunday school; but she wouldn't think of it. Doesn't Ralph look cute in his little Buster Brown suit? Molly made it herself." Bessie didn't really talk loud, but her voice was clear and rang in the little room.

There were ten pupils in the school—I was going to say, children, but I couldn't, because Rudolph Haas was nearly as tall as Father. He was in the eighth grade. They all watched us like chicken hawks while Miss Wheeler had us read and do numbers for her. After we were done with the numbers, she decided Grace belonged in the fourth grade and I in the third.

Recess didn't go a bit good for me. Before we left home, Mother had taken us into the front room and said,

"I am not going to have you children grow up to be rowdies and ruffians just because we live on a farm. Ralph, if you get into a fight in this new school, I shall give you a hard thrashing when you come home. The Bible says that if your enemy smites you on one cheek, you are to turn the other. I want you to follow that teaching absolutely. And Grace, I want you to promise me on your word of honor that you will tell me if he ever raises a hand against any other child at school."

She must have heard about Freddie Sprague. He was in the second grade, but he was bigger and fatter than I. We hadn't been out to recess a minute when Freddie put a stick on my shoulder and then knocked it off. "Wanta fight?" he said.

Grace was standing right behind him, and hollered, "If you do, I'll tell Mother."

I knew she would, too, so I said, "No, my mother won't let me."

I don't know why the Bible picked out cheeks, but that's right where Freddie hit me. I wanted to hit him back, but I didn't dare to. Mother could spank pretty hard if I did something right after she'd told me not to. I just let my hands stay down and turned my face around. Freddie hit that side, too. And he hit it hard.

When I started to cry, somebody sang out, "Molly, Molly." Then all the boys, and even some of the girls, started yelling it.

Grace came over and wiped my face with my handkerchief. "Don't cry," she said. "Mother will be proud of you." Then Miss Wheeler rang the bell for us to come in, and said Freddie should be ashamed for hitting me.

Every time she wasn't looking the rest of the morning, he kept making faces at me.

Noon was worse than recess. Grace had brought a rubber ball to school with her. She knew how to play jackstones with it, but the other girls didn't. As soon as we got out at noon, she divided our lunch and went off with the other girls to hunt for the right-sized stones. Mother had baked us each a cup custard. I laid my two sandwiches and piece of cake on top of my cup and went to look for a place to eat it. When I was halfway out to the carriage shed, Freddie came running with another stick to put on my shoulder. I spilled my sandwiches and cake trying to get away, and he knocked the cup out of my hands, yelling, "Fight, Molly!" All the other boys laughed at me. Some of them yelled, "Hit the little sissy, Freddie, make him fight." I had to run away to the far corner of the yard.

All the other youngsters either rode to school or drove in old buggies. There was an open shed where they backed the buggies in at one end and tied the horses at the other. Willie Aldivote rode a funny-looking grayish-white donkey with big yellow splotches. After they had eaten their lunches, the other boys got the donkey out and started to ride him. They would hold him by the ears until some boy got on, then they'd let him loose and tickle him with a switch. He would buck like sixty. Sometimes the boys stayed on and sometimes they fell off.

I had come back to the corner of the shed where I could watch. When Freddie saw me, he hollered, "Let's put Molly on him," and they did.

I wasn't very much afraid. I hadn't fallen off the cowboy's horse, and thought I could cut the mustard. Anyway, I was willing to take the chance. Mother hadn't said I couldn't ride a donkey, and I wanted to show the boys I wasn't a sissy. My ride lasted about half of the donkey's first buck jump. He kicked his heels up, and I pitched off between his ears. I landed right on my face in the gravel. I skinned my nose and bit my tongue. I couldn't help crying a little, but I didn't make any noise.

Mother nearly went frantic when we got home from school. My Buster Brown suit was pretty well messed up from the bloody nose I got when I fell off the donkey, and Freddie had given me a good working over after school. I guess it hadn't helped my looks very much. Grace was so busy telling Mother about my being a good boy and not fighting that, until after I was all cleaned up, she forgot to give her the note Miss Wheeler had sent.

Mother's eyes got all full of tears when she was washing my face and putting court plaster on my nose. She told me that she was proud of me for being a little gentleman, and that she would see to it that those ruffians didn't attack me again. Grace hadn't seen me try to ride the donkey, and I thought it would be best not to mention it to Mother.

Father came in from digging post holes just after Grace had remembered about the note. Mother was madder than ever and read it aloud to him. The teacher said that Grace had adjusted herself very nicely with the other girls and she was sure I would soon make my adjustment, but she would suggest that I wear overalls to school like the other boys.

Mother tossed the note down on the table and slapped it with her hand. Her lips were pinched together and she said, "The fact that I am going to bring my children up on a farm isn't going to prevent me from bringing them up to be ladies and gentlemen. Ralph wearing overalls to school! The very idea! I shall see that teacher the first thing in the morning."

Father shut his eyes and scratched the back of his head. Then he said, "There might be some reason in what she says, Mame. And, you know, there have been some gentlemen in overalls." Mother's lips pinched tighter than ever, and she never said a word.

Father and Mother sat up in the kitchen till after I went to sleep. I don't know what they talked about, but the part that was about me must have ended in a compromise. Mother didn't go to see the teacher, and I didn't wear overalls to school.

My time at school wasn't happy. Freddie Sprague made it miserable, and I got so I was scared to death of the yellow-spotted donkey. Freddie thought it was a lot of fun to grab me by the back of my pants and jerk hard enough to pull off a couple of buttons. Then I'd have to hold them up with one hand until I could get to Grace for a safety-pinning. Mother always made them so I had to unbutton them when the other boys didn't have to, and some of the older ones were always yelling, "Come, Molly, let Mama unbutton you." Or they'd run me down and lug me for another nose dive off the donkey.

One raw, windy afternoon when we were at recess, Willie Aldivote made an I-betcha that he could hold his breath longer than any of the rest of us. Nobody won because I turned blue and fell over after half a minute. It didn't help me a bit at school, but it really worked fine at home.

Though it wasn't nearly so cold in Colorado as it had been in New England, the high altitude bothered my heart a little, and I felt the cold more than the others. Once or twice I got so cold on the way home from school that I turned good and blue. When I did, Mother would doctor me with brandy and sugar. I tried for it often, because I liked it, but she would only give it to me if I was really blue to the ears.

I found that if I stayed outside long enough to get good and cold, then held my breath for half a minute just before I staggered through the kitchen door, it would get me a spoonful of brandy. Mother never guessed, but after I had worked it half a dozen times, Father caught on and I became what our old minister used to call a teetotaler.

Father tried to make up to me a little for the trouble I was having at school. He picked Saturdays to go to Denver for loads of lime and secondhand bricks to build a foundation under the house. I always went with him, and each time we had a picnic of custard pie and milk. Then, on the way home, he let me drive Bill and Nig.

One evening after school, when Father and I were working on the foundation, Fred Aultland drove into our yard. He and Father talked for a while about the house and building the foundation, then Fred said, "Charlie, you seem to be pretty handy with the tools. You know, my father used to run cattle all through this section when I was a kid. He kept eight or ten cowhands all the time, and sometimes there'd be as many as a couple of dozen. The old bunkhouse hasn't been used for ten years or more. I been thinking about tearing it down, but all I can get out of it is firewood. I'll make a deal with you. If you want to buy me a ton of coal and haul it out from Denver, I'll take it in trade for the bunkhouse."

It was fun moving the bunkhouse. We did it one Saturday. It was bigger than our whole house, and I couldn't see how in the world we could move it without tearing it all to pieces, but Fred and Father figured it out. Fred and his two hired men helped us, and it was easy as pie. Father shaped the ends of four double-length fence posts so they'd fit into the hubs of the wagon wheels. Then he bored holes through the ends of them and put axle grease on the shaped part. As the men pried up one corner of the bunkhouse I'd lay cribbing sticks under, the way Father showed me. We just kept going around and around it; they lifting a little and I putting in another stick or two, till it was high enough to run the shaped ends of the posts under it and into the hubs of the wheels. When the stay pins were driven, it made the whole bunkhouse into a great big, eight-wheeled wagon. Of course, we couldn't turn any sharp corners, but we didn't need to. We just hitched on four horses and hauled it down across the prairie.

At our house we took it off the same way we had put it on. Father was fussy about making it land L-shaped to the back of our house, with the corners just touching.

Although I was having a lot more fun at home, things

were getting worse and worse at school. The Friday after we moved the bunkhouse, it got so bad I couldn't stand it any longer. We had just finished eating our lunches that noon when Freddie Sprague started picking on me. He yanked at my pants so hard that all but one of the buttons flew off, and I had to use both hands to hold them up. Then he got dirty and yelled to the other kids, "Let's pull Molly's pants off so he won't have to squat like a girl." They did, and right where all the girls could see, too.

I didn't care whether Mother would be ashamed of me or not. I couldn't be a gentleman with my pants off, and I didn't want to be one anyway. I plowed into Freddie with both fists. I had a big advantage, because he didn't expect it. His nose started bleeding before mine did, and that made me sure I could lick him. He tried to quit after a few minutes, but the older boys wouldn't let him out of the ring. I had been dreaming about that day for a month, and kept hitting him as fast as I could make my arms go.

I heard the girls' voices around the ring and looked right up into Grace's face. She was jumping up and down and yelling, "Hit him! Hit him! Hit him hard!" I did my best, and Freddie put both arms up over his face. Once when the boys were pulling him back onto his feet, I glanced up toward the schoolhouse. Miss Wheeler was peeking out the corner of the window, but she didn't ring the bell till it was all over and Grace had pinned my pants back on.

Rudolph Haas lived way over south of the schoolhouse, but that night he drove Grace and me clear to our house in his buggy. Just before we got in, Miss Wheeler came and said for Grace to tell Mother that she thought I had made my adjustment now.

Maybe she thought so, but Mother didn't. She started to cry as soon as she saw me—my eye had turned kind of greenish-black. She even cried while she was paddling me. It was a good, hard spanking with one of Father's slippers, but I didn't mind it at all; it was worth *two* spankings to lick Freddie. I don't remember all the things she said to me, but they were plenty. Then she told Grace to call Father. He sent me for the milk as soon as he came into the house.

The next day was Saturday, and I worked with Father all day. He was cutting the bunkhouse in two, so we could move one of the pieces around and build it into a new kitchen. Along in the middle of the forenoon, I was holding a window sash while he took off the frame. He had his back to me, and we hadn't been talking at all. All at once he said, "I hear you had a fight yesterday."

I had been expecting it. I didn't look up, and he didn't turn around. I said, "Yes, sir."

"Did you lick him?"

"Yes, sir."

"Good." That was all. He never mentioned it again.

4

MY CHARACTER-HOUSE

WHILE WE WERE DOING OUR LESSONS THAT NIGHT, Mother said, "Didn't Fred Aultland offer to lend you horses until ours were well again?"

It was a few minutes before Father answered. "Yes,

Mame, he did, but . . . what is that one you quote about being neither a borrower nor a lender? Fred's a fine fellow, but I don't want to start borrowing from the neighbors, and then, too, he'll need all his horses at plowing time when I'll have to have three."

Mother said, "I just don't see how we can take out fifty dollars for another horse, and more to buy harness for it, until we have something to sell." Father didn't say anything except that he'd talk to Mr. Cash.

Mr. Cash was an old horse trader–peddler, who used to come by our place every couple of weeks. He had an old covered wagon filled with everything from safety pins to rolls of secondhand barbed wire, and he usually had three or four old horses tied to the tail gate.

He came by the next Sunday, and Father haggled with him for over an hour. I was out by the wagon when they finally made the deal. Father had looked the different horses over from teeth to hoofs. He had both hands in the hip pockets of his overalls and was looking at an old bay mare on the off side. "How much for Grandma?" he asked.

Mr. Cash said he had paid twenty-five dollars for her, but in the end he let Father have her for seventeen-fifty, and threw in a collar, and hames with chain traces. When we were leading her to the barn, I noticed that she dragged one fore hoof. Father said it was because she had been foundered. Because she walked so slow, Mother named her Nancy Hanks, after a famous race horse.

Mother let me wear my overalls and blue shirt to school

Monday morning. Grace fiddled along all the way, hunting for some certain kind of cactus. We got in just after the bell rang. Freddie's eye was blacker than mine and his lips were still puffed. I had wanted to get there early enough to see if I could lick him again before school began, but I had to wait for morning recess.

Everybody knew Freddie and I were going to fight again at recess time. We made faces at each other every time Miss Wheeler had her back to the class. What I didn't know was that Freddie had a deal with Johnnie Alder to help him lick me.

Johnnie was in the fifth grade, and nearly as big as Willie Aldivote. As soon as we got out near the carriage shed, Freddie and Johnnie both jumped on me. I was sure I was done for until Willie sailed in, too.

I always liked Willie Aldivote best of the boys. He was the second biggest in school; he was twelve, and was in the sixth grade in number work. He was the one who had the yellow-spotted donkey, and he could ride like everything. He had never been mean to me.

Willie told Johnnie, "If you're going to help Freddie, I'll help Molly," and the fight was on. It was really two fights for a little while. Then, when Johnnie was good and licked, Willie kept Freddie from holding me by the hair with one hand while he punched me with the other.

After that I had better luck, and before the bell rang for the end of recess, Freddie had both arms up over his face again. Then he did something that I ought to have loved him for. He said, "There ain't meat enough on him to hit, that's why I can't lick him. He's just a sackful of old spikes, and I hurt my own hands worse'n

38

I hurt him every time I punch."

At noon I ate my lunch with all the rest of the fellows, sitting right between Willie Aldivote and Rudolph Haas. After we finished eating, Willie said, "Come on, Rudy, let's teach Spikes how to really ride that jackass." Nobody will ever write a poem that will sound as good to me as that did. I knew I wasn't going to be Molly any more, and I'd have tried to ride a rhinoceros if they'd said so.

They got straps off one of the harnesses and a piece of good, stout cord. After they'd buckled the straps real tight around the donkey, Rudy held him by the ears while Willie put me on and tied my feet together with the cord. He ran it snug between them, under the donkey's belly. Willie told me not to be scared, to squeeze tight with my knees and watch the donkey's head so I could lean way back when he put it down to kick. I grabbed hold of the straps with both hands, and Rudy let him go.

For what seemed an hour, but was probably ten seconds, things were happening too fast for me to keep up with them. I tried to watch his head and lean back every time it went down, but it was bobbing so fast I lost the beat. My head snapped back and forth as though it were tied on with rubber, and I bit my tongue. My head rang so I couldn't even hear Grace yelling—she had come running to pick up the pieces when I got killed—but I was still on top when he stopped, and I'd had fun. After school, Willie let me ride behind him on the donkey as far as our corner, and Grace was so proud of me that she promised not to tell Mother.

Three or four days after we got Nancy Hanks, Grace

and I came home from school through our west field, so I could show her where Bill and Nig had fallen off the trestle. There were twelve or thirteen long crossties in the bottom of the gulch, and I told Grace that Father had said for me to drag them home as soon as we got another horse. I had been riding Willie's donkey every day, and thought I'd be able to handle any horse.

I guess I'd made up too many stories before. Grace didn't believe me, and told Mother about it the minute we got home. I should have admitted to Mother right then that it was only a story, but I was sure she'd spank me if I did. The more I thought about hauling those ties with Nancy, the more I wanted to do it. And, since I was going to get a spanking anyway, I thought I might just as well have the fun of trying. So I swore up and down that Father did tell me. He had gone to Denver after another load of bricks, and I was sure I could have a crosstie home before he got back. Then he would probably be proud enough of me so that I wouldn't get spanked at all.

Mother had to help me put the collar and hames on Nancy, because they were heavy and the collar didn't have any buckle at the top. You had to put it over her head upside-down, and then turn it around. I never heard Father criticize Mother, and that was the only time I ever heard her criticize him. She put her hands on her hips and buttoned her mouth up tight, then she said, "I don't know what in the world this ranching has done to your father! Insisting that you wear overalls to school and be permitted to behave yourself like a guttersnipe! And now! Hmfff! Sending a little eight-year-old boy off to haul logs with a new, untried horse!"

40

The crossties didn't haul as easy as I thought they would. I forgot to take anything along for hitching them to Nancy's singletree, and had to use an old piece of barbed wire. I was walking on the downhill side of the crosstie when we tried to go up over the bank at the head of the gulch. Suddenly the tie started to roll toward me and I had to dive out of the way. I skinned my nose and the barbed wire tore a big hole in my overalls, but the tie missed me, and Nancy seemed glad to stop and rest. I was still trying to haul it up over the bank when it got dark, and I was afraid Father would get home before I did. It would be better to go right home and tell Mother I'd lied. If she spanked me first, he probably wouldn't give me another one right on top of it.

Mother didn't spank me, though. She gasped, and looked at me as if I'd been a rattlesnake ready to strike. Then she made me stand with my face in a corner. I stood there while the rest of the youngsters had their supper and went to bed, and for at least an hour afterward while Mother sat at the table and read the Bible. I knew what she was reading, because I heard her take it down from the lamp shelf. There wasn't any clock in the kitchen, and the only sound was the thumping of my own heart.

At last I heard Father drive into the yard, and listened to every sound as he put the team away and came to the house. He stood in the open doorway for a moment before he spoke. When he did, his voice was very quiet. "What has happened, Mame?"

It was a full minute before Mother said anything. And then her voice was as quiet as Father's. "Charles, the

41

time has come when this boy must have a father's firm hand. I am appalled by the degeneracy he has shown since we left East Rochester."

I had never heard Mother's voice like that, and I had never heard her call Father "Charles." I thought my heart would pound itself to pieces while she was telling him what I had done. Hard as Father could spank, he never hurt me so much with a stick as he did when Mother stopped talking. He cleared his throat, and then he didn't make a sound for at least two full minutes.

When he spoke, his voice was deep and dry, and I knew he must have been coughing a lot on the way home. "Son, there is no question but what the thing you have done today deserves severe punishment. You might have killed yourself or the horse, but much worse than that, you have injured your own character. A man's character is like his house. If he tears boards off his house and burns them to keep himself warm and comfortable, his house soon becomes a ruin. If he tells lies to be able to do the things he shouldn't do but wants to, his character will soon become a ruin. A man with a ruined character is a shame on the face of the earth."

He waited until his words had plenty of time to soak in, then he said, "I might give you a hard thrashing; if I did, you would possibly remember the thrashing longer than you would remember about the injury you have done yourself. I am not going to do it. There were eighteen crossties in the gulch yesterday, and the section foreman told me they were going to replace twenty more. Until you have dragged every one of those ties home, you will wear your Buster Brown suit to school, and I will not take you anywhere with me."

It was half a mile from the house to the gulch. Father showed me how to hook onto the ties with a chain, and how to pull them up through the head of the gulch. By getting up early, I dragged one tie home each morning and two after school. With a half dozen on Saturdays, I had the job done in a couple of weeks.

It was a tough two weeks. I was sure I would become Molly again for wearing the Buster Brown suit to school; I think Grace saved that by telling everybody what a bad boy I was. Then, too, the weather turned windy and cold. The gulch half filled with dry tumbleweeds that scratched me when I dug through them to get the ties. But worst of all, Mother got a song in her head. When that happened, she would sing the same tune over and over and over, for a week at a time. That time it was "The Bird with a Broken Pinion." I don't think she was singing it just for me, but I couldn't go into the house without hearing about "a young life broken by sin's seductive art," or that "the soul that sin hath stricken never soars so high again." It made me think a lot, as I walked along behind Nancy and a dragging crosstie, about the permanent damage I had already done to my character.

5

THE BIG WIND

I THINK IT WAS MY LYING ABOUT THE TIES THAT GOT US the buckboard. One night I came in from tie hauling to hear Father and Mother talking together in the kitchen.

Mother was saying, "But, Charlie, we can't just load the children into that old farm wagon like cordwood and haul them off to church. We saved quite a little from what we expected to pay for the new horse and harness. Somewhere we should be able to find something better in the way of a conveyance that's within our means."

Mr. Cash came by on Saturday, and he had our buckboard tied behind his wagon. Sunday we all went to church at Fort Logan.

Church and Sunday school were held in the day-school room. There weren't enough seats to have them both at one time, so the grown folks stayed outside and talked while we had Sunday school, and we played outside while they had church. The fathers of most of the other children who went to Sunday school were soldiers at the Fort. Some of the kids were kind of tough. I learned a couple of new words from them, and Philip learned a lot.

He called Muriel a couple of them after we got home, and Mother cried so hard that Father sent us all into the bedroom. We could only hear a word or two she was saying between sobs. It was something about not being able to stand what the ranch was doing to her boys. From then on Mother held Sunday services at home.

Father had been hauling long poles to be cut into fence posts for about a week when the big wind came. It was blowing when we woke up, and tumbleweeds were rolling across the prairie like big brown bowling balls pitched by some giant in the mountains. By school time it was too strong for Grace and me to stand against. By

44

noon it had racked our house until some of the windows had broken out and the doors were jammed fast in their casings. The whole house was vibrating like a beaten drum, and every few minutes a joint among the rafters would crack with a report as sharp as a rifle shot.

Father's face was gray, and Mother's milk-white. Neither of them spoke; their mouths were clamped tight, and the muscles were popping out and in on the sides of Father's jaws. I could see Muriel, Philip, and Hal crying, but, against the roar of the wind, I couldn't hear them.

Father went out and untied the horses. They drifted away to the east, the wind whipping their tails up under their bellies. Next he brought poles and propped them against the lee side of the house. Mother huddled us into a corner of the bedroom, away from the windows. She crouched over us like a hen brooding her chicks. There came a tearing screech from the roof as the wind ripped away a section of the shingles, and sheets of plaster fell from the ceiling.

Father crawled in through the blown-out window with a coil of rope in his hand. He took his Sunday suit from the corner and told Mother to put it on. Then he knotted the rope around our chests and shoulders until all but Hal were strung on it the way Mother used to string popcorn balls for Christmas—about five feet apart. Philip was on one end and Muriel on the other. Mother had taken off her dress and put on Father's suit, with the sleeves and legs rolled way up. Father tied Philip's end of the rope around her waist and Muriel's end around his own.

Then he motioned Mother to follow, tied Hal on his back like a papoose, and crawled out the window. As she passed us out to him, he had us fall to the ground and lie still. After he lifted Mother down, he crouched and told us to crawl on our stomachs like horned toads; that dust would get in our eyes, but we must keep them open so as not to crawl into cactus beds.

Our nearest neighbors were the Aultlands, a mile up wind. Fort Logan was to the east—three and a half miles away—with no houses between. Father crawled east, and we crawled after him.

When we had wriggled along for a hundred yards or so, Father stopped to let us rest, and I looked back toward Philip and Mother. Philip must have gotten cactus spines in his hand, because he held it out toward Mother and tried to sit up. The wind caught him and rolled him like a tumbleweed as far as the rope would let him go. As he went, Mother sprang to her hands and knees. She was no more than up before she sprawled forward on her face as though some giant had put his foot on her from behind and shoved. In that same backward glance, I saw the roof of our new barn fly away like a sheet of newspaper. We started on. The next time we stopped to rest, I looked back again. There was blood on Mother's face, and our barn was gone completely.

A few minutes after we had begun crawling again, something like the shadow of a great bird flashed past me on the ground. I raised my head, and a second later the body of our farm wagon struck a few feet beyond Father. It bounced crazily like a football and flew away in kindling wood.

My eyes were running from the dirt in them, my nose burned as though the dust in it were pepper, and I was

46

coughing from breathing through my mouth. At the next rest, I lifted my head again and looked up and down the line. Father was coughing hard—I could see Hal bounce up and down on his back. Philip was sobbing and gasping for breath against the pull of the wind, and Mother's face was black where dirt had mixed with the blood.

I had no idea where Father was taking us, but after a dozen or more stops I knew we were going more north than east; we were not going to Fort Logan. We crawled across the wagon road, and on, and on. The wind ripped up curled, dry leaves of buffalo grass and raked them across our faces like jigsaw blades. At last Father stopped and waved his lifted arm. Then he raised them both and made motions like a man pulling himself up a rope. We all understood and drew ourselves up to him. There was fresh blood at the corners of his mouth. We got our heads close to his, and he yelled, "We're almost there. We're going to be all right."

It seemed hours longer before we crawled into the head of a gulch leading down to Bear Creek. Under the level of the prairie we could crawl on our hands and knees. Father led us along to where there was an over-hang in the west bank of the gulch. There was hardly any wind there. I was so tired I could barely move, and shaking all over. I wasn't frightened any longer and nothing was hurting me, but I started to cry. I didn't know what I was crying for, but I couldn't make myself stop. Everybody but Father and Mother was crying.

The wind went down with the sun. At twilight it was no more than a breeze. We crawled from our shelter under the bank, cold, stiff, and ragged. Our faces were smeared with blood, cut by the sharp grass that had been whipped against us by the wind, and our hands, arms, and legs were burning with cactus spines. It seemed to me that we had crawled miles on our stomachs, but when we came out at the head of the gulch, our house stood little more than half a mile away. It and the bunkhouse stood alone; barn, privy, wagon, and buckboard were gone.

The house leaned tiredly against the prop poles. From the wagon road it looked like a deserted ruin. And when we got to the back door Father stood Hal down and put his shoulder against the door. It stuck tight and he heaved against it, but not as he had heaved when he lifted Nig's hind legs from the trestle. He looked tired and old.

The inside of the house looked almost as deserted as the outside. Fallen plaster, broken glass, and dirt covered everything. Father coughed hard after pushing the door open, and wiped his mouth with a red-stained handkerchief.

Sometimes Mother cried over little things, but she didn't cry then—she hadn't all day. She bustled right through the kitchen and into the bedroom, with her underlip bitten in between her teeth. In a few seconds I heard her shaking bedclothes so hard they snapped. While she was doing it, Father looked over the chimney to see if it was cracked, then started a fire in the stove. Next, he took boards from the bunkhouse and nailed them over the broken kitchen window. We youngsters didn't know where to begin in helping get cleaned up, and stood huddled by the stove.

When Father had finished with the window, he put his hat on and started out the door. Mother came from the bedroom while he was in the doorway and said, "Charlie, where are you going?" She spoke quietly, but her voice was husky—way down in her throat.

Father said, "I've got to find the horses, Mame. Heaven only knows where they may be by now. They may have fallen in a gulch."

Mother went over and took hold of his arm. She stood close against him and looked up into his face. Her voice was still husky as she said, "Ralph can stay home from school tomorrow and find them; you're going to bed."

Father bent over and kissed her on the forehead as he tried to take her hand off his arm, but she wouldn't let go. "Charlie," she said—it was only a little more than a whisper—"we came here to save your life. Are you going to throw it away over so little? We need you, oh, we need you, Charlie." From where we were standing, I saw her eyes fill up with tears, but none spilled over.

As I watched her, I heard the fast beat of running horses' feet. Fred and Bessie Aultland came into the yard, circled, and pulled up at the steps. Fred jumped off the buckboard with the reins still in his hand, and cried, "Charlie! For God's sake, what's happened to you, man? You look like a ghost."

After Mother told them about the house nearly blowing over, and about our crawling to the gulch, Bessie said they had worried all day because they knew we didn't have a storm cellar. They made us all go home with them and stay for three days, while Fred and Bessie helped Mother get our house fixed up again. Mrs. Ault-

49

land and Mother wouldn't let Father get out of bed for the whole three days.

6

We Become Real Ranchers

It rained most of the next couple of weeks. One of Fred's hired men had found our horses and what was left of the wagon and buckboard. Nig was all right, but Bill stood with his back humped up most of the time, and all the starch seemed to have drained out of Nancy's legs. We kept them in the end of the bunkhouse Father had cut off to use for a kitchen, while he and I worked in the other end. There was no school when it rained, so I could stay home and help him a lot. We built a new wagon body and put new spokes in the wheels of the buckboard, right there in the bunkhouse. Between showers we built a new privy—we were never able to find even one board of our old one.

I had to bring two pails of milk from Aultland's every night. When I had eggs to bring, too, Grace went with me to lug the basket. Mother was making Father drink all of one bucket of milk every day, and giving him raw eggs with a spoonful of brandy in the glass. I still liked the smell of that brandy, but the weather was getting warm enough so that I couldn't get blue any more. Father's cough got better day by day, but as he grew stronger, Nancy grew weaker. One morning near the end of the second week, Father went out to the bunkhouse and found her dead. Grace and I had to go to school that day, so we missed the funeral.

Our big wind must have been an ill one, because I never heard of anyone it did any good—except maybe it did help us a little in one way. The Saturday after Nancy died, Father and I were putting the buckboard back together when a man with a big load of new boards stopped out in front of our house. He came in and said his name was Wright, and that he lived a couple of miles up the creek. He said the wind had raised the dickens with his buildings and he'd noticed how nicely we were getting fixed up again.

After he'd watched Father work for a little while, he said, "You seem to be a pretty handy sort of fellow with a hammer and saw. I wish I could get you to come and help me get fixed up. I'd give you three dollars a day for your time, or trade work with you, or trade something I might have that you wanted."

Father told Mr. Wright he ought to be getting started with his own plowing now that the rains had come, but he would talk it over with Mother and let him know the next day. Mother must have said it was all right, because Father helped Mr. Wright for a couple of weeks.

Father drove Bill and Nig up to Mr. Wright's the first morning. When he came home he had a saddled bay mare tied to the tail gate of the wagon, and a little black and white collie puppy in the pocket of his reefer coat. He said Mr. Wright had insisted on lending him the mare to ride back and forth, but the puppy was ours to keep. It was still so young it was wobbly on its legs and cute as could be. We all wanted to claim it, but Father wouldn't let us. He said Muriel could name it, but it would belong to all of us. She named it King.

I liked King a lot, but it was the saddle horse that really took my eye. She was a crabby Morgan mare and laid her ears back every time anyone went near her. In the mornings when Father put the saddle on, she would fling her head around and snap at him with her teeth, but she always missed him by an inch or two. She could run like a greyhound, and I wanted to ride on her so much I couldn't think about anything else, but neither Father nor Mother would let me go within a rod of her. I had got so I could ride Willie Aldivote's donkey without my feet being tied together—and only holding on to the belly strap with one hand. Willie had taught me how to squeeze my knees tight behind the withers, and ride on them instead of the seat of my pants. I had spent hours practicing and knew that, especially with a saddle, I could ride the mare as easy as pie.

It was about the middle of April when Father finished helping Mr. Wright repair his buildings. He took Bill and Nig with him that last Friday. I remember so well because that was the day we changed from being immigrants to being ranchers. When he came home that night the bay mare was tied to the tail gate. She didn't have the saddle on, but there was a driving harness in the wagon, along with four little Berkshire pigs and two gunny sacks with the heads of half a dozen hens sticking out of holes in each of them.

I wanted to claim and name the new mare, but Father wouldn't let me. He said we had all, except Philip, named something, so he must have his choice; he chose Fanny.

Mother didn't want the end of the bunkhouse as a kitchen, after the horses had been living in it, so Saturday morning we hooked Bill and Nig to it and hauled it out where the first barn had stood. It had been built with board walls inside and out, and the space between them stuffed with straw. We worked to beat the band all day. After we got it moved Father ripped up the floor and pulled off all the inside boards. He let Grace and me pull the old nails and pound them out straight, while Muriel and Philip were lugging the straw to the hen coop and pig pen he built with the floor boards.

When Father hauled the piece for the barn away, it left the bunkhouse with one end open. That Sunday he built a new end into it with boards he had pulled off the inside of the barn, and made a partition in the middle, so there was a room for Philip and me, and one for Grace and Muriel. While he was doing it, Grace and I helped Mother move the beds and make us bureaus out of boxes the groceries had come in. She put cloth around them—with ruffles—and we made scalloped paper covers for them, and cut doilies from pieces of old wallpaper. By supper time everything was done, and Grace and I were so excited about sleeping in a real bunkhouse that we could hardly get away from the table quickly enough.

Father had promised Mother that he would plow her a garden out behind the barn before he did anything else. He started it early Monday morning, but he hadn't got around the plot once before we had to go to school. He had made a tripletree for the plow, and balanced it carefully so to adjust the amount of pull to the strength of each horse. Nig was to walk in the furrow and pull the biggest share, Bill in the middle with the next biggest, and Fanny on the outside with only a little more than half as much load as Nig.

But Fanny had no intention of being a plow horse. She was all right while Father was putting Nancy's collar and the hames with chain traces on her, but when he tried to rein her up beside the other horses, she squealed and tried to bite chunks out of Bill's neck. Father fixed that quick enough by fastening a stick about three feet long between her bridle and Bill's collar, but when he hooked her traces, she kicked and jumped around till she had both hind legs over them and was faced the wrong way. Father unhooked her and talked easy till he had her back where she belonged, but every time he got the traces hooked to the tripletree, she would do the same thing all over again.

If it had been Mother, I think she would have killed Fanny right then and there, but Father didn't seem to get mad at all—only the muscles on his jaws went out and in. At last he made another jockey-stick and put her in the middle with one stick fastened to Bill's collar and the other to Nig's. That way she couldn't swing out around, but she did kick like fury, and got both legs over the traces. Bill and Nig didn't get any more excited than Father did while Fanny slatted and threw herself around. She acted just like a little kid in a tantrum.

When she had quieted down a little, Father put the reins over his shoulder, stuck the point of the plow in the ground, and clucked to the team. Bill and Nig started up when Father clucked, but Fanny stood stock still. The tripletree caught her on the heels, and then the real fun started. Fanny went away from there like a stone out of a slingshot. When she reached the end of the jockey-sticks, she went straight up, and came down bucking, with heels flying in all directions. Finally she got one foreleg inside Bill's bridle rein and one hind leg inside Nig's breeching, then she went down with Nig on top of her. When Father got them untangled, he made me go to school. I never heard Father swear, but I always wondered if he didn't that time as soon as I was out of earshot.

I stopped by Aultland's for the milk on my way home from school. Fred was out by the corral fixing a disk harrow, and I went over to tell him how Fanny had acted when Father hooked her up to the plow. He laughed as though it were a big joke. "It's no wonder old man Wright traded her off cheap," he said. "He's spent ten years and a dozen sets of harness trying to break her to drive double. I'll sure take my hat off to your old man if he can plow half an acre with her. Why the hell wouldn't the stubborn down-east Yankee let me lend him a good horse to plow with? Say, how much land is he figuring to turn over this year, anyway?"

When I told him Father was going to plow the whole place if Bill held out, he squinted up one eye for a minute, and said, "Go on in and get your milk; I'll give you a lift home."

Father had most of the garden plowed when we got there. The big horses were walking slowly, just one step after another. Fanny was soaking wet and tossing her head up and down, but she was plowing, so I told Fred he'd have to take his hat off to Father.

Father put his foot up on the hub of Fred's buckboard the way he always did. They talked about what would be best to plant on new sod ground. They talked and talked. Then Fred said, "Charlie, how much of this place do you figure on putting into crops?"

Father looked over toward the horses and said, "I'd like to put it all in, Fred, but with the late start I've got, and at the rate I've been going today, I guess I'll be lucky if I get in eighty acres."

Fred just sat chewing for a minute or two, then he squirted a line of tobacco juice between the nigh horse's heels. "You know this prairie land won't produce much in the way of grain crops the first year, and drinks up a hell of a lot of water. A fellow ought to put in crops like peas and beans and alfalfa the first year, so's to get air back into the land. Why don't you put in about ten acres of alfalfa? We've had quite a bit of rain this spring, and if you sow it with oats—and get it in before the first of May—it might get roots down to moisture before it burns out on you. Then you could put in another ten to peas and beans, and you'd have about all you wanted to take care of this first year."

Father stood looking down at his foot on the hub of the buckboard for all of two minutes, then he looked up at Fred and his voice was real quiet when he said, "What are you telling me, Fred—haven't I got any water?"

Fred didn't answer till he'd spit between the off horse's feet and cut another corner of his plug. "Yep, Charlie, you've got water—ten inches. This land will produce forty bushels of wheat to the acre if you've got an inch of water to the acre. Without an inch to the acre, you're lucky if you get any."

Father pushed his hat back and scratched his head a little. "Can I count on getting the full ten inches, Fred?" he asked.

"That's the hell of it," Fred said. "You're tail-ender on the ditch. When the creek's high and the ditch is running full at the dam you'll get your share, but when it's running low and the crops are burning up, you'll be able to lug all you get in a bucket. I won't steal water from you, Charlie, but when only half my own is coming through to me and my crop's suffering, I won't pass it on to you."

Neither of them said anything for a long time, then Fred said, "Your cousin ought to have found out about it before he got you out here. Why, man, you couldn't run ten inches of water to this garden from where the ditch comes onto your place; the ground would drink it all up on the way. I'll tell you what I'll do. I've got two hundred inches with my place. I'll use all the water that comes as far as me for twenty days, then give you the whole head for one. That'll let you give about twenty acres a good soaking often enough to make a crop the first year. After that you might handle as much as twenty-five."

7

I BECOME A HORSEMAN

FATHER AND MOTHER MUST HAVE SAT UP AND TALKED nearly all of that night. I woke just as the moon was slipping down behind the mountains, and there was still a light burning in the kitchen. Mother had brought some garden seed from New England and had bought more at Fort Logan. The next day she let me stay home from school and help her plant peas and potatoes and carrots and beets. We dug trenches most of the forenoon, then Mother sent me to shovel the horse manure from behind the barn onto the wagon, so Father could haul it out for us at noon when he came in from plowing.

Mother had me put manure in the bottom of the trenches and cover it over with an inch or two of dirt, then she laid in the cut pieces of potato and hoed dirt over them. We were right in the middle of it when I looked up and saw half a dozen cowboys riding by on the wagon road. I waved, and one of them turned his horse and came cantering across the prairie to where we were. I knew him as soon as he got near enough for me to see his face. He was the same cowboy who had given me the ride.

He flipped out of the saddle while his horse was sliding to a stop, and took his hat off to Mother with a sort of half bow. "I see you folks are really gettin' dug in. We was scairt the big wind might have blowed you clean out of the country."

While he was talking to Mother I was looking at his horse. It was a blue roan, the first one I had ever seen. "Yes, we're here to stay," Mother said. "My husband is going to build a storm cellar, so there won't be any danger of our being blown clean away."

I wished Mother hadn't said "clean away." It sounded the way she did when she didn't like somebody, and I wanted her to like my cowboy friend. I walked around the roan and looked at him from the other side while Mother and my cowboy kept talking.

Mother didn't talk much, but the cowboy said, "Lady, you're sure wastin' your time buryin' these here barn cleanin's under your spuds; you're due to get tops enough without it. All you got to have for this ground is water, and God help the man that ain't got it."

The hair on the blue horse was shinier than it was on Cousin Phil's Prince. It rippled like oily water when he moved the muscles under it. To me, it was like a magnet. I had to touch it with my hand, so I stepped up close to his shoulder. Just as I reached my hand up, the cowboy called, "Hey, Pardner, watch out, you're on the off side. Come on around here."

While I was coming around, he said to Mother, "This old cayuse is clever as a kitten if you stay on the nigh side, but he might kick the stuffin' out of him over on the off side." He had ground-tied the roan by just dropping the reins when he got off. He picked them up while he was talking and passed them around the horse's neck, then he caught me by one arm and swung me into the saddle. "How about a little ride, Puncher?" he asked.

Mother thought he was just going to lead the horse around a little with me on it, and she didn't say anything except "Be careful," when he was showing me how to stick my feet into the loops of strap that held the stirrups. As soon as I got them in, he passed me the lines and clucked. The roan went off in a smooth, easy canter, and Mother cried, "No! No! He'll fall!"

My friend laughed, and I could hear him say, "Aw shucks, if he falls, the ground'll catch him."

It didn't. At first I held on to the saddle horn with one hand—the ground seemed so much farther away than it did when I was riding the donkey—but I didn't feel a bit as though I were going to fall off, so I let go and waved back to Mother and my cowboy. The only time I was frightened at all was when I went to turn him around to go back. We had gone clear out by the railroad and I was afraid he might fall down going across, so I pulled on the left rein, but he swung around to the right. For just a second I thought I was going to take a header, but I kicked out hard with my left foot and was back in balance again.

I could see Mother was peeved when we came cantering in; her mouth was pinched up that way. For just a second I thought about seeing if I could flip off as my friend did when he came, but the ground was a long way down and I was scared to try it. Anyway, my feet were stuck in the loops. He reached up and took me off while Mother stood with her hands on her hips. She looked at me with that look of hers that said, "Come here, young man!" And I went. She didn't say a word to me, but her eyes blazed at the cowboy as if she would like to skin him. "You might have killed him," she said. "If he'd fallen off, that horse would have trampled him to death."

He just laughed, "No, Ma'am! That old pony wouldn't kill nobody. If he'd a fell off, Old Blue would of just stood there and waited for him to pick hisself up. You watch."

Then he said to me, "Didn't have no trouble with him, did ya, Little Britches?"

I said, "No, only he don't steer very good. I pulled the

rein to make him go one way and he went the other."

He laughed again. "He's just rein wise and you ain't, that's all. Now you watch."

Then he turned around toward Mother, took his big hat off and said, "You watch, too, Ma'am, and you'll see how safe he is."

He kicked up one leg and flew right into the saddle without ever touching the stirrup. He whistled between his teeth as he went up, and Blue was gone with his feet kicking chunks of sod out behind him. The roan had hardly gone fifty feet before he sat right down on his hind legs and skidded, then the cowboy made him do more tricks than an organ grinder's monkey. They turned round and round in a circle, and from one side to the other so that it looked like dancing, then he would run a little way full tilt and be turned around before he got through sliding. I noticed that the cowboy never did pull on either rein; he just held them in his left hand up over the horse's neck, and whichever way he moved his hand, that was the way the roan went. Then he did one that made Mother and me both squeal. With the pony going lickety-larrup the cowboy fell right out of the saddle. He lit on the back of his shoulders, turned a half somersault and came up on his feet. The horse stopped so fast they were standing there side by side, as if they were just waiting for the mailman to come along.

The cowboy looked around at Mother and took off his hat. It had stayed on all the way through the somersault. He stepped back into the saddle again and trotted over to where we were. First he said to me, "Catch on, Little Britches?" Then he took off his hat to Mother again, and

said, "Hiram Beckman's the name—they call me Hi."
As he raced back toward the road he turned and waved
his hat. Mother and I waved back.

I could hardly wait for Father to come in from the field
to tell him about Hi and his blue roan horse. Father had
been plowing way over across the tracks, and I didn't
think he'd noticed us, because he never stopped to look
when I could see him. I ran out to meet him when he
came, and got all mixed up, I was trying to tell him so
fast. He put his hand out and rumpled up my hair. I didn't
know what he meant, but he said, "I guess you're a chip
off the old chopping block. If you understand them, you
never have any trouble making them understand you.
You did all right on that horse. I knew you weren't afraid
by the way he was acting." We walked along a little way,
then he rumpled my hair again and said, "Your father
was proud of you, Son." It was the first time he ever told
me that, and I got a lump in my throat.

Then he told me that Hi might be a little bit of a show-
off, but he was a good horseman; not so much because
he could fall off and come up on his feet, but because he
had been patient in training Blue. He said that Blue
wasn't a bit afraid of Hi or he wouldn't have handled so
smoothly, and that it was the best example he had ever
seen of complete understanding between a man and a
horse. "If you want to be a good horseman," he said, "the
first thing you'll have to learn will be how a horse thinks,
and next to think the same way yourself."

That Sunday was nice and warm. After the chores were
done, Father said, "Mame, this is too nice a day to be

cooped up in the house. If Fanny hadn't been plowing all week, I'd say let's hitch her up to the buckboard and take a drive up to the mountains, but she hasn't steadied down yet, and is making twice as much work of it as she needs to. So, what do you say—let's pack up a picnic lunch and a good book, and make a day of it down by the creek?"

We all went running around trying to help Mother get ready faster, when we'd have helped more by keeping out from under her feet. By ten o'clock the big lunch basket we had on the train was packed, and we were on our way down over the hill to Bear Creek. Father found a place where the creek made a wide curve through a grove of cottonwood trees and tumbled down in a cascade to a deep, clear pool, lined with willows. He showed us how to skip flat stones on the pool, and then we all went wading in the creek—even Mother and Hal. Mother took a puckering string from her petticoat, and a safety pin, so Philip could go fishing in the pool, while Father taught me how to whittle a willow stick into a whistle.

Grace and Muriel went up the creek to pick up colored stones while Mother unpacked the lunch basket and boiled water to make tea for herself and Father. Pretty soon Grace came running back, calling for us all to come quick, she'd found a whole bushel of pure gold and had left Muriel to guard it till we got there. We all went running but Father. He tried to act as if he were hardly interested, but he did walk faster than usual. All the way, Grace kept babbling about how we were rich now and could get a cow, and a pony to drive to school. When we got to where Muriel was, the sand near the shore was all covered with shiny yellow flakes. Father took some of it on his hand and looked at it carefully. Then he said, "Girlie, I wish you were right, but I believe it's mica. I think they call it fool's gold. I read about it once, but if I hadn't, I'd certainly be fooled, too."

After we had our picnic, Mother read to us. She didn't read like other people; she talked a book. I mean, if you were where you could hear her but couldn't see her, you'd be sure she was telling the story from memory instead of reading. And another thing different about Mother's reading was that she didn't care if you watched the book over her shoulder. I used to watch her eyes by the hour as she read. They would swoop across the page like a barn swallow across a hayfield, then she would look up and recite for a full minute before she looked back at the book again. When Mother read, we children had to be quiet and pay attention. We could do most anything we pleased with our hands, like making whistles, stringing dried berries for beads, or playing with dolls, but if one of us whispered, Father would snap his fingers. If he ever got to the third snap, Mother would close the book and we would do something else for a while.

I don't remember Mother ever reading anything I couldn't understand, and I never heard any of the others say so either, but I don't think many people would have read us the same books she did. That day it was *John Halifax, Gentleman*. Maybe she skipped spots we couldn't have understood, and maybe some of it drifted over our heads, but at least we remembered the stories she read. I think part of the reason was that we could raise a hand whenever we wanted an explanation of any word or situation.

I liked John Halifax a lot, but as the afternoon passed, I found my mind wandering from the tannery to the open range, where Hi might be punching cattle on his blue roan. The more I thought of Hi, the farther I left John behind. After Mother had explained to our Muriel Joy that Father took her name from that very book, I suggested that maybe I should leave early, to get the milk before it was too late. I had my plans all made, if Father said yes. He said it.

I started up over the hill in the direction of Aultland's, but as soon as I got over the shoulder of the first rise of ground, I headed for home as fast as I could scramble. I got the milk pail—a ten-pound lard bucket—and set it in the wagon. Next, I untied Fanny's halter rope and led her out there, too. I tied her to one of the wheels, with less than a foot slack in the rope, so she couldn't back away. Then I got her bridle, took one of the long reins from the driving harness, and fastened an end to each bit ring. By standing in the wagon box, I could reach her head all right, but I was afraid she would run away when I took her halter off, so first I tied the loose end of the rope good and tight around her neck. Fanny was one of those mares that fought the bit, but I didn't know it, nor what to do about it. I guess I just expected her to open her mouth wide and wait for me to lay the bit into it. When I showed her the bridle, she tossed her head and pulled back to the end of the rope. I leaned out of the wagon as far as I dared, holding the bit up toward her lips. When I got it close, she would bob her head up and down and swing around where I couldn't reach her.

As Fanny kept dancing away from the bridle, I kept one eye peeled for sight of the folks coming back from the creek. Usually, we would beg Mother to say poetry for us after she had stopped reading. Sometimes we could keep her going for an hour or so, but I was usually the one who did most of the begging. If they had got Mother going on a good, long one like "Horatius at the Bridge," I'd be all right, but if it was just a short one like "The Day is Done," I was sunk.

The more Fanny jerked her head around, the madder I got at her and the more afraid I was that I would get caught before I had a chance to try to ride her. I climbed out astraddle of the wheel and tried to push the bit in between her clenched teeth as her head bobbed. Finally I remembered that Father talked quietly to her when he made her plow, and decided to try it. I got down and patted her on the shoulder. As soon as her ears were pointed forward, I untied the halter rope and pulled it up easily till I had her chin right up to the wheel tire. Then I tied it tight and climbed back on the wagon. I kept telling her what a nice mare she was as I offered her the bit again. It made no impression; she still kept her teeth locked.

My time was running out. Even if it was Horatius, it couldn't last forever. I stuck one thumb in between her lips and gouged down with my thumb nail. That seemed to be something Fanny understood. She opened her teeth and took the bit. I was so excited I forgot to buckle the cheek strap, but grabbed up my bucket and shinned over onto her neck. When I had worked my way back to the withers, I untied her neck-rope, and we were on our way.

I was quite surprised to find that she was easier to ride than the kicking donkey—and her withers were slim enough so that I could get a good knee hold.

Fanny didn't canter smoothly like the blue roan, and I didn't have any stirrup straps to balance myself with, but I was still on top when we got to Aultland's. I tied her way over at the end of the pole corral, hoping no one would see I had ridden her—and so that I would have the poles to climb up onto when I was ready to get on again.

I hadn't fooled anybody at Aultland's. I guess they had seen me coming up the road. Fred said, "I knew your paw was proud about you riding Hi's pony, but I'd a bet your maw wouldn't let you try Wright's mare bareback."

I remembered what he'd said before about betting his life I'd make a horseman—and I thought maybe if I acted like Mother knew already, they wouldn't bring the matter up later—so I said, "Oh, she saw me ride Hi's blue horse and she knows I'm going to be a horseman. She doesn't care."

Bessie made me take the milk in their can instead of in our lard bucket, and it was lucky she did. I tried to hold the can still on the way home, but Fanny seemed to be in a hurry to get there and ran like all get-out. The can got bouncing up and down and I was so busy holding on with my knees that I just had to let it bounce. The cover popped off and milk went everywhere. As we came out of the last gulch, there was the whole family coming up over the hill from the creek. The nerves in my bottom started to tingle—and it wasn't from rubbing on Fanny's back.

It was too late to turn back. I knew Father would have

seen Fanny, because her feet were clattering on the adobe road like sticks on a snare drum. For just about a second I thought he might not have seen that I was on her, and that I might be able to dive off like Hi and come up on my feet. The ground was going by so fast that I was actually afraid to look down, let alone dive off, but I didn't want to admit it, and told myself I'd better not try it because I'd spill the rest of the milk.

I had planned to ride Fanny up to the wagon, so I could get off without dropping the milk can, but she had her own ideas about where she was headed for, and shot right into the barn. I could see I was going to be raked off if I didn't do something about it, and do it in a hurry. I dived head first at the manure pile, milk can and all. That's where Father found me when he came running around the corner of the barn a minute later. I wasn't hurt a bit, and I still had the empty milk can, but my best Buster Brown was kind of messed up.

Mother and the rest of the youngsters were only seconds behind Father. Mother was furious after she got over being scared, and demanded that Father give me a good, hard spanking. She said he could talk to me till he was black in the face, and it wouldn't do a bit of good, because my wickedness was so great that it had killed my conscience. Nothing but fear of bodily pain would save me from a life of crime.

Father didn't say a word, but just turned me over his knee—while I was trying to tell him that I hadn't lied to be able to do what I wanted to, so hadn't injured my character any more. Father had a trick I never knew about before. He must have cupped his hand up some way, because every whack sounded like it was killing me, but they hardly stung at all. I howled loud enough to make up the difference.

8

I BECOME A SORT OF COWBOY

BY THE BEGINNING OF MAY, SCHOOL HAD PRETTY WELL petered out. Nobody sent boys to school when they were needed at home to help with the plowing or planting, so when it got down to where only four girls and I were left, school closed for the summer.

The day after it closed, Mrs. Corcoran came to see Mother about getting me to work for them. They had about thirty milch cows, and used to take cream to Fort Logan every day. In the summer they pastured the cows on the quarter section south of us. Because there weren't any fences, somebody had to herd them to keep them from getting into Aultland's and Carl Henry's grain fields. She said she would pay me twenty-five cents a day, and I would only have to work from seven in the morning till six at night. I guess Mother thought they herded cows on foot in Colorado, as they did in New England, so she said I could do it.

I didn't have any such ideas at all and was all excited about being a cowboy. My biggest worry was that I didn't have a ten-gallon felt hat, instead of a straw one from the grocery store at Fort Logan. I spent the rest of the afternoon out behind the barn, twisting myself a pair of spurs out of pieces of baling wire. It seemed best to sort of take it for granted that I was going to ride Fanny, so I was up and dressed in time to help Father feed the horses before breakfast. I shoveled manure to beat the band while he was bringing in the hay, and said, "Which bridle had I better put on Fanny for herding cows?"

Father grinned at me and rumpled up my hair—I had combed it that morning with plenty of water and had made a hook in the front lock so it hung down over my forehead like Hi's did. He said, "Sorry, Son, but I guess you'll have to take it on your feet today. Carl's going to let me use his drill to plant the alfalfa, and I'll have to use Fanny."

I nearly hopped up and down. I tried to keep my face straight, but I was laughing all over inside. I was sure from the way Father said it that he was going to let me ride Fanny, now that he knew I could do it. It didn't make any difference if I had to wait a day or two. After breakfast I got my spurs from where I had hidden them under the hay, and poked them in the front of my blouse. I thought I'd be at least part cowboy if I had spurs, even if I did have to walk.

Mr. Corcoran was a milk-cow man and not a horseman. He didn't have any nice horses like Fred Ault-land's bays, or Carl Henry's chestnuts. They were mostly horses about like our Bill and Nig. There was an old black plug in the corral with the cows. He had faded out to a brownish color. Mr. Corcoran brought out an old work-harness bridle with blinders and put it on the horse. Then he boosted me on and gave me a switch. "Old Ned ain't too spry, but you give him a cut with that switch and he'll get a move on. Now don't let none of them cows

70

get into Fred's or Carl's grain or they'll skin you alive. And don't run none of the cows—some's with calf and they're all milkers."

I wasn't too happy with Ned, but at least he was a horse. I had driven the herd nearly as far as the wagon road when Mr. Corcoran bellowed after me, "Be careful not to let 'em get no green alfalfa, it would bloat 'em and kill 'em."

Everything went fine till I got past Aultland's house. Fred's field, from the house to the section corner, was unfenced and half a mile long—and it was in alfalfa about six inches high. My cows spied it from a hundred yards away, and some of them started running for it. I kicked Ned with my heels, but he wasn't at all nervous, and didn't even hurry his walk. Then I clipped him a little with the switch and he took half a dozen trotting steps before he went back into a walk. His feet were as big as footballs, and every time he trotted, I bounced a foot high and came down with a thud. He was a lot wider in the withers than Fanny, so I couldn't get a good knee clamp on him, and I wasn't a bit sure I wasn't going to bounce clear off his back.

Some of my cows had already reached the alfalfa, and I expected to see them start falling over in great, bloated corpses. I swung my switch high and started cutting it down over Ned's rump—my spurs, which I had twisted onto my bare feet as soon as we reached the road, had crumpled at my first kick. At the second cut, Ned got the idea I wanted him to hurry and trotted a dozen or so more steps. I was so busy staying on that I couldn't think to swat him again and lift him into a canter. Now all the cows were in the alfalfa, and I knew my career as a cowboy was blowing up right in my face. I didn't care about falling off any longer. They had to be gotten out of there some way. After a couple more hard licks the switch broke in my hand. There was only one thing left to do, so I piled off Ned and took after the cows afoot, yelling at the top of my lungs. The only stick I could find was too heavy for me to handle with one hand, so I waded into the herd swinging it like a baseball bat. Instead of driving them back into the road, I only drove them farther into the alfalfa field.

I was so busy swinging and yelling that I didn't see Fred until his tall bay horse was almost on top of me. Fred had a long blacksnake whip and snaked those cows out of there in about a minute and a half. Ned was making the most of his chance. He hadn't moved a foot from where I slid off, and had his nose buried in the alfalfa halfway to his eyes. Fred told me to go get him while he kept the cows moving. I couldn't much more than reach Ned's belly, there was nothing to climb up on, and I had no idea how I'd get aboard him. Fred yelled, "Hang over his neck and kick; he'll hist you up."

He did, and I went up with my club in my hand. Ned had a lot more respect for it than he did for the switch, and I caught up to Fred in a hurry. The things he said about Mr. Corcoran were good to listen to. They were just the things I would have said myself if I hadn't been afraid of the damage it might do that character of mine— I wished Father had never told me about it.

Fred helped me till we got the cows over onto the piece of prairie where I was supposed to pasture them, then he gave me his blacksnake and told me not to be afraid to lay it on if I had to. I had forgotten all about my spurs, but Fred saw them and laughed. He said that baling wire was the only thing that had held the State of Colorado together, but he'd bet I was the first one who ever made spurs out of it. Before he left he showed me how to swing the blacksnake so as to make the cracker pop right behind a cow, and said to let Ned have the handle across the rump if he wouldn't go. Then he told me to try to keep the cows bunched pretty well in the middle of the quarter section, and that he'd have one of his men come to help me take them home at night.

It was a terrible day. The quarter section wasn't flat like our place, but was all rolly hills. And those cows knew more tricks than Hi's blue roan. As soon as Fred was out of sight, they started spreading out in all directions. I beat the tar out of Ned, trying to make him go fast enough so that I could keep them rounded up, but the most I could get out of him was that clumping trot. While I was driving back a few stragglers on one side, others would head for Carl Henry's oat field on the run. When I came back with the first bunch that tried it, I found that I only had nineteen cows left in the herd—the rest had got away over one of the hills.

I beat on Ned's rump and went to hunt them. His trot was pounding the dickens out of my behind and it was getting awful tender. I had worn off a piece of skin the size of a silver dollar. Ned had started to sweat a little right where I was trying to sit, and the salt in the sweat made it sting like blazes. When I got over the hill, I saw my strays a quarter of a mile away, headed for the oat

73

field. They were in it before I could catch up to them.

Father must have been watching me from where he was sowing alfalfa. I had left Ned and was wading around in the oats, trying to drive the cows out with the blacksnake. I couldn't handle it very well when I was on horseback, but it was almost useless in the oat field. I didn't have strength enough to keep it in the air through the back swing, and the cracker got tangled in the oats. I guess my yells were getting sort of warbly, and I was about ready to cry when Father showed up on Fanny.

He got them out of there in no time, helped me to collect the rest of the herd, and bunched them way over at the east end of the quarter. Four more times he had to come to my rescue before six o'clock, and then he helped me get them back to Corcoran's. After we had the cows in their corral, Mrs. Corcoran came out from the house and tried to give Father my quarter. He nodded his head over toward me and said, "Give it to the boy, he certainly earned it."

I don't think Mrs. Corcoran liked what Father said, because her face got a little red. She passed the quarter up to me and said, "Now don't go and lose it the first thing you do." Then she said to Father, "Too much money ain't good for children. These young ones nowadays haven't no idea of the worth of a dollar. I don't know what things are coming to."

Father only said, "Better slide off him, Son. Mother'll be waiting supper for us."

I slid off and put Ned in the corral with the cows. All the time it took me to climb on the gate to get his bridle off, Mrs. Corcoran kept talking. First she said, "Little boy, you didn't let my cows get into nobody's crops, did you?"

I kept looking right at the cheek strap buckle, but I knew I had to tell her, so I said, "Well, sometimes I couldn't make Ned run fast enough to—"

That's as far as I got. She sounded mad as could be. "My land sakes alive! You ain't been running my milch cows all over creation, have you?" I tried to tell her I hadn't, but she didn't even stop to breathe. "Good gracious!" she said. "I'll wager I don't get more'n half a milking tonight. How a body's going to eke out a living from half milkings, I just don't know. Well, hmmmf, I suppose some allowance has to be made on account of him being city-raised. City-raised young ones ain't been learned to do things when they was young. They don't have the gumption of them that's raised in the country."

Father reached his hand down and pulled me up back of him on Fanny. As he swung me up, he said, "We better be getting along, Son."

I knew I was fired and got an ache in my throat. It had been a tough job, and I hadn't done very well, but I had been counting all day on the time Father would let me ride Fanny. I was sure I could manage all right with her, and now that I was fired from my first cowboy job, I was afraid Father would never let me ride her.

We were pretty near out to the road when Mrs. Corcoran yelled after us, "You be sure you ain't late in the morning—right sharp on seven o'clock." Raspy as her voice was, it sounded good to me.

Fanny could canter right along with Father and me on her. Sitting way back where I was, I couldn't get a knee hold, so I had to put my fingers under Father's belt. I held as easy as I could, so he wouldn't notice and think I was afraid of falling off. We were about halfway home when he said, "It's a pretty big job for a city-raised fellow; want to take another crack at it, or have you had enough cows?"

I said, "I could do it all right if I only had Fanny."

"Well, I guess I could spare her tomorrow," Father said—that was all.

I don't think Father ever told Mother what Mrs. Corcoran said about city-raised young ones, because they kept right on being friends. When we got home, she let me put my quarter up in the new cupboard, in her Wedgwood sugar bowl. She knew about Father having to come over and help me, so when he came in from feeding the horses, she said, "Charlie, don't you think that is a job for a man, not for a boy of Ralph's age?"

Father grinned, "They're certainly a breachy lot, but I have an idea he can make out. There's one old heifer up there that I don't think he could handle, but he won't have to ride herd on her."

I couldn't figure out which one he meant, but I guess Mother knew, because she looked at Father out of the corner of her eye, and said, "Charlie, now you behave."

Maybe my day's work didn't please Mrs. Corcoran very much, but it made me quite a hero with the other youngsters at home. That quarter was the first money any of us had earned, and it looked as big to them as it did to me. After the supper dishes were done—I didn't have to help with them now I was a working man—Grace got a pad of paper and a pencil. First she asked

Father how much a cow would cost, and then she wanted to know how much it would take for a pony and a cart. She put them all down and added them up, then she divided the total by twenty-five cents. When she had the answer, she went for a calendar, but before she could do any more we had to find out if Mother would let me work on Sundays. We knew that would be a ticklish job, but Grace figured out the way to do it.

She said, "Mother, is it sinful to cook on Sunday?"

Mother was busy sewing a thick pad into the seat of my new flour-sack underpants. Father had told her about my getting the skin worn off on old Ned. She looked up, and said, "Why no, of course not. God made us so that our bodies need food on Sunday the same as any other day. Since He has just loaned us these bodies for the time we are here on earth, it is our responsibility to take the best care we can of them, so there is nothing sinful about preparing food for ourselves on Sunday. But what put any such question into your mind?"

"Oh, I was just thinking. Mother, did God lend cows their bodies, too?"

Mother didn't look up, but said, "Yes, dear."

"Well, Mother, would it be sinful to feed cows on Sunday?"

That time Mother did look up. "Why, of course it wouldn't be sinful; it would be cruel not to feed them on Sunday. What in the world makes you ask such a silly question?"

"Well . . . if it wouldn't be sinful to feed them on Sunday, it wouldn't be sinful to herd them on Sunday either, would it? I was just wondering if you were going

77

to let Ralph herd the Corcorans' cows on Sundays."

Mother jabbed the needle down into the pad and looked up frowning. "Most certainly not!" she said. "Mr. Corcoran can feed his cows hay on Sunday. Now put that pad away and bring me the Bible. Ralph has to go to bed early and get his rest."

Father had been reading a farm magazine. When Grace started asking Mother about cooking on Sunday, he let it down enough so he could look over the top, but as soon as she said, "Most certainly not," he boosted it right up again. I couldn't see how his face looked, but he had wrinkles at the corners of his eyes just before he lifted it.

9

GRACE TRIES IT, TOO

I GOT AWAY FROM HOME ON FANNY A LITTLE AFTER half-past six the next morning. Father must have sat up kind of late the night before, because he had made me a new sort of whip for the cows. It was the end of a broom handle about a foot long with a piece of harness rein fastened to one end and a rawhide loop to the other. The piece of rein leather was five or six feet long, with the end sliced into four narrow strips. It was a lot lighter than the blacksnake Fred had lent me. Father showed me how to put the rawhide loop around my arm and snap the stick so that I could hit things with the split end of the rein.

I didn't take any lunch that day. Before, Mother put me

up sandwiches and cake, but I tied the package on my overall strap and lost it while I was chasing cows in the alfalfa. She said Grace would bring me a hot dinner at noon, and watch the cows while I ate it.

I was scared to death when I took the cows out into the road. I knew they would run for Fred Aultland's alfalfa as soon as we got past his house, and I was pretty sure I couldn't keep them out alone. That was before I knew much about Fanny. The Corcorans had one spotted brown and white cow that was skinny as an old hound dog. She was always way out in front of all the rest, and she could run like a horse. We were hardly past Aultland's driveway before she started running for the alfalfa, and I knew the others would follow her unless I could head her off. I was just thinking about going after her, and I guess maybe I leaned forward a little bit. Anyway, before I clucked to her, or kicked my heels, or anything, Fanny was after her on a dead run. She almost went right out from under me, she started so quick, but I grabbed hold of her mane and stayed on.

We caught up to the spotted cow before she got halfway to the alfalfa. I planned to slow Fanny down beyond her and get turned around so I could drive her back with the others. Fanny hadn't planned it that way. When her head was a foot or two in front of the cow's, she and the cow both turned—all in a second. I was the only one that didn't. It all happened so fast that I never remembered hitting the ground. I scrambled to my feet, scared that Fanny would run for the barn at home. She didn't, but stood and let me get hold of her rein. The old cow was running back to the herd as fast as she had run away from it.

The ground was as flat as a table, and I was panicky for fear I couldn't get back on Fanny before the cows were all in the alfalfa. Then I remembered how Fred told me to climb on Ned's neck the day before, but first I had to get Fanny's head down. I ran over to the side of the road, yanked up a handful of grass, and held it out toward her nose. When she started to nibble, I dropped it in the road and threw myself on her neck as soon as she put her head down for it.

I took half a dozen more spills before we reached the pasture, but none of them hurt very much. Fanny knew all the tricks there were about making cows do what she wanted them to, and my biggest job was guessing which way she was going to turn, and when. And all the way there were fields of alfalfa or oats along one side of the road, so I could climb back on her neck when she put her head down to eat. Just before we turned into the pasture, I filled the front of my blouse with green oats. I knew I'd fall off some more, and I had to have a way of making Fanny put her head down.

Fanny was much easier to ride than Ned—even if she did spill me once in a while. The only time she ever took a trotting step was when she was slowing down to a walk after cantering. She could canter along as slow as old Ned trotted, or she could go like a streak of greased lightning. I found out that the farther I leaned over her neck, the faster she would go, and maybe I ran her fast lots of times when I didn't need to.

Grace brought my lunch at noon. It was "everything stew" in a lard pail, and biscuits and a cup cake. When she brought it, my cows had wandered nearly to the south end of the pasture, so there were a couple of hills between us and our house. Grace said Mother had told her to herd the cows while I ate, and she wanted me to bend over so she could use my back for a stepping-stone to get on Fanny. I tried to tell her she didn't know how to ride and would fall off, but she got kind of mean and made me do it. She knew a couple of things about my fighting at school and riding on the donkey that I didn't want her to talk about at home. She didn't really say she'd tell if I didn't help her get on Fanny, but she did remind me that she hadn't yet.

I told her about clamping her knees and watching Fanny's ears. I was getting so I could tell when she was going to turn and which way, because she would point her ears that way first. Just as I got the lid off the lard pail, my old spotted cow started toward Carl's oat field at a trot. I yelled to Grace to head her off, and Fanny acted as if she knew exactly what I had said. She went racing off after the old cow as fast as she could go. Grace was almost lying down on Fanny's neck, and her bottom slewed way over to one side. I knew she wasn't squeezing with her knees, and yelled to her. It was too late.

Fanny caught up to the cow, and Grace wasn't watching her ears. How she ever fell as she did, I'll never know. She was clinging to Fanny's neck with both arms and had dropped the reins—I had them tied together so they wouldn't fall if I let go of them. When Fanny turned so quick, it swung Grace out like a gate, and her feet came down between Fanny's forelegs, but she was still holding on with her arms. Fanny kept right on going until she had the old cow headed back, then she stopped and just stood still. By the time I got over there, Grace was standing on the ground—laughing and crying all at the same time.

Grace had heard Willie Aldivote tell me that if you fell off you had to get right back on and try again, else you'd be too scared to try later—and besides the horse would know you were scared and you could never ride that one again. I knew Grace was frightened silly to get back on Fanny, because she was shivering as if it were the middle of winter, but she wasn't going to let me be able to do something she couldn't, so she made me bend over again while she stood on my back. That time she didn't act so smart when I reminded her about pinching her knees and sitting up straight and watching Fanny's ears. I told her her hands weren't very stout yet, just as Fred Aultland told me, and showed her how to wrap the reins around them.

She must have been even more scared than I thought she was. I started her going away from where the cows were, so Fanny wouldn't see some old heifer she thought ought to be chased. As soon as she moved one foot, Grace pulled up hard on the reins and Fanny stopped. I clucked to her, but Grace pulled harder and yelled at me to keep quiet. Her pulling and yelling made Fanny cranky, and she began bobbing her head as she did when she didn't want to plow. Then she started going backward so fast she was almost sitting down. I yelled to Grace to let up on the reins, but I don't think she heard me. She grabbed hold of Fanny's mane with her right hand, so that rein went loose, but she kept on pulling with the other hand. Fanny began going around in a circle backwards, and I didn't know what to tell Grace to do. I guess we were both yelling as loud as we could, and the louder we hollered the faster Fanny went around.

Father always used to say the worst things you expected never happened to you. That's the way it worked with Fanny. I didn't dare tell Grace to slide off for fear Fanny would step on her, and I guess she didn't dare to either. When I thought she was a goner for sure, she fell forward and hugged Fanny around the neck again. As soon as both reins went slack Fanny stopped, and I ran in and got hold of her bridle. Grace was glad enough to call it a day's ride, and even bent over to let me climb on. It would have been easier to shin up Fanny's neck as I usually did, because Grace's back was wobbling around like a patted dog's. After I had the cows rounded up again, she herded them on foot while I finished my dinner. Then she took the bucket and started for home, but when she got to the top of the first hill she yelled back to me, "I can ride better than you can any old day. I can ride her going backwards and you can't." I didn't even bother to answer her.

I was afraid Grace might have ruined Fanny, but she didn't. I only fell off once all afternoon. But I thought I was sunk that once, because I had run all out of green oats to make her put her head down. I had planned to get some more while Grace was watching the cows at noon, but her getting in such a mess with Fanny made me forget all about it. I pulled a handful of dry buffalo grass and held it out to her, but she wouldn't even sniff it.

When I had my mind all made up that I was going to have to lead her clear over to the oat patch, she hung her head down and I scrambled on. From that time on, Fanny and I had an understanding between us: if I fell off she'd put her head down for me to get on again, but if I got off by myself I had to get back on the best way I could.

I had a little trouble getting the cows home that night. Leaving the pasture, about half of them streaked off ahead toward Carl's oat field, while the rest dragged along behind. I went kiting after the leaders, and while I was getting them headed off, the others got past me by running up a little valley where I couldn't see them. Fanny and I got them out easy enough, but by that time the first bunch was back into the field a hundred yards or so farther down the road. We raced back and forth between the two herds till Fanny was in a lather, but as soon as I got one herd out, the other was in. Carl's house was beyond a hill, so he couldn't see me, but we were right in plain sight of Aultland's. I kept looking to see if Fred wasn't coming to help me again, but he didn't. At last I woke up to the fact that all I had to do to get them all out was to let one herd stay in till I could drive the other up to join them, then drive them all out together.

We got by Fred's alfalfa all right, and I was proud as I could be that I hadn't had to have any help all day long. I was still being proud of myself when Mrs. Corcoran came out with my quarter. She had a safety pin, too. Instead of giving me the quarter in my hand, she put it into the pocket of my blouse and safety-pinned it in. I left it there till I got clear out to the road, on my way to

Aultland's for our milk. Then I took it out and put it in my overall pocket, so I could feel more like a man. But I stopped Fanny in the bottom of the last draw before we got to our house and pinned it back into my blouse pocket. I couldn't be sure Mrs. Corcoran and Mother hadn't cooked the idea up between them.

Fanny was pretty sweaty when I got home that night, and Father didn't like it. He told me I was wearing her down because I hadn't learned to make my head save her heels. I made the excuse about the two different bunches of cows getting into the oats and how hard I had to ride to get them out, but Father said, "Now wait a minute, Son. Every time you've been in sight all day, you've been playing cowboy, haven't you?"

Of course, I had been, but I didn't know how Father knew. I nodded my head. "Do you want to be a good cowboy like Hi," he asked, "or do you want to play at being a cowboy?"

"Like Hi," I said.

"Then spare your horse. A cowboy with a spent horse is in as bad a spot as if he didn't have any horse at all. Hi wouldn't waste his horse's strength any more than your mother would waste our money—that is, not unless he was showing him off for her benefit. Instead of racing around after every cow that strayed a few yards from the herd, he'd put them all at the back end of the pasture where he could see them from the top of a hill. Then he'd sit down and let his horse graze until some of his cows had wandered far enough away that they might get into the oats. When he did have to go after them, he wouldn't race as you do. He'd go at a nice easy lope till he was past the strays, then bring them back at a slow walk so as to keep them calm and quiet. Always remember, Son, the best boss is the one who bosses the least. Whether it's cattle, or horses, or men; the least government is the best government."

The next day went pretty fine for me. I only tumbled off Fanny once, and I wouldn't have had to that time if I'd grabbed hold of her mane. Once, the day before, I had got off balance and knew I was going to fall, so I let go of the lines and reached my hands out to catch myself on the ground. I came down smack on my face and nearly broke my arms. This time, we were right in the middle of a sandy spot at the bottom of a little valley. I had been studying all morning about the way Hi fell out of his saddle on purpose and somersaulted onto his feet, so I thought I'd try it. As I went off, I ducked my head and bucked up my hind end. It worked, but it worked too well. I went too far over in the air and came down on the seat of my pants with an awful thud. The sand wasn't half so soft as it looked, but at least I'd learned part of the trick of taking a fall.

That morning I herded the cows the way Father had told me Hi would do it. They seemed to know I had learned the trick, and I only had to go after them two or three times. The rest of the morning I kept right on top of a hill where Father could see me from our bean field. But when I saw Grace coming with my dinner I moved down into the little valley with the sandy spot.

I wanted to show her that I could fall off on purpose without getting hurt, and that I was brave enough to do it with Fanny galloping. I thought maybe I could do the somersault trick so I'd come right up on my feet. It didn't work any good at all. There must have been a big old jack rabbit that I didn't see, sitting right at the edge of the sandy patch. I had Fanny going like sixty and had loosened up my knees, all ready to take my dive, when she set her feet and stopped dead still. I went off over her head a mile a minute. If I'd gone a couple of feet farther, I could have grabbed the old rabbit as he raced away.

It happened too fast for me to think anything about any fancy landing, and I made a perfect belly slide. It knocked the wind out of me for a second. When Grace got there I was all right, but I couldn't get any air into my lungs so I could say so. She dropped my dinner bucket and came screaming like she thought I was killed. I don't think Fanny liked her very well after the day before, and she shied away. I was afraid she might run home before I could get breath enough to yell "whoa" at her, but she didn't.

My dinner was a mess. Mother had put the baked beans in the bottom of the bucket, then put a saucer on top of them with my johnnycake and pie on it. When Grace dropped the bucket it all got mixed together—it was lemon pie, too. All the time I was eating, Grace kept telling me that it was her duty to tell Mother about my falling off Fanny. I begged her not to, because I knew Mother wouldn't let me ride any more if Grace ever did tell. At last she said she wouldn't squeal, even if it was going to hurt her conscience, but I'd have to help her get on so she could ride Fanny. She promised she wouldn't haul on the lines.

Grace got on all right, but I kept hold of the reins till I

saw she was sitting right and had her knees squeezed in good and tight. Then I held Fanny's bridle and talked to her easy till Grace got the lines wrapped around both hands. Grace was all right as long as I had hold, but when I let go she leaned forward and grabbed for Fanny's mane. The minute she leaned forward Fanny started to canter. Grace squealed, and I hollered after her to sit up straight and keep the reins tighter, but not to haul on them. She did sit up, but she hauled on the lines.

I don't know whether Fanny was trying to be mean, or whether she didn't know what Grace wanted her to do—and I don't think Grace knew herself. Anyway, she started trotting right up the little valley. Grace went bouncing up and down on her back like a marble dropped on a stone walk. It wouldn't have been so bad if she had just come down in the same place every time, but sometimes she was clear up on Fanny's withers, and sometimes pretty near back to her tail. First she'd lose her balance one way, then she'd grab a handful of mane and pull herself half off the other side. Why she never fell clear off I'll never know, but she didn't. At last she got worked way up on Fanny's neck, and slipped over sideways so far that she was just hanging by her hands and one knee. Then Fanny stopped and let me catch up to them. Even at that, Grace yelled back to me when she got to the top of the hill with my dinner bucket, "I guess I showed you who could ride best. You fell off and I didn't."

Grace brought my dinner every noon, and she always had something hurting her conscience enough so that she'd have to tell Mother if I didn't let her ride Fanny.

After a while I just let her do it anyway, and she got so she could do pretty well, but she was always a sissy, because when she found she was going to fall she'd grab Fanny around the neck. After a day or so Fanny'd stop as soon as Grace started hugging her.

I got so I could tumble into the sandy spot and hardly get hurt at all. And a few times I went clear over and came up on my feet like Hi. I didn't have any trouble with the cows after the first week. When June came, the days were hotter and I didn't have enough to do for it to be interesting any more. Mrs. Corcoran stopped hollering so much about my running the cows or bringing them in too early, but she still pinned my quarter into my blouse pocket every night, and I always took it out and put it in my overall pocket till I was nearly home.

10

MY FRIEND TWO DOG

WHEN I GOT IN FROM HERDING ONE EVENING EARLY IN June, there were two scrawny, jug-headed buckskins tied to an old rickety spring wagon in our yard. I could see them from half a mile up the road, and came boiling home a lot faster than Father liked me to. Nobody was in sight, so we tore right for the barn. Fanny and I had a system at the barn. If I stayed on she had to go through the doorway slow or I'd get scraped off. But she always liked to pop right in quick, so I'd slide way back on her rump and slip off over her tail at the last second. I'd got so I could do it and land on my feet most every night—

without spilling the can of milk I always brought from Aultland's.

I had to time it just right, and that night I was thinking so much about who might be in the house that I must have been a little careless. Anyway, I hadn't slid back far enough on Fanny's rump, so when I ooched to go off over her tail I didn't go all the way, and crashed into the header of the barn door. When I woke up I didn't have any idea where in the world I was for about a minute. I was lying on a pile of hay and an old man with a long white beard and a battered ten-gallon felt hat was looking down at me from squinty blue eyes that were sunk way back in his head. I shut my eyes again quick, and heard him say to somebody else, "Ain't hurt a bit, ain't hurt a bit. 'Fraid the little papoose mighta brained hisself."

I remembered what had happened then, and knew where I was. The first thing I thought about was the milk, because Mother had been giving me heck every time I spilled some of it, so I said, "Did I spill the milk?"

The old man laughed and laughed, then he said, "No Bucko, you didn't spill scarcely none of it. Two Dog ketched it soon as ever it lit."

I sat up then and looked around to see who Two Dog was. He was a wizened old, old Indian, and his face was so wrinkled it looked like a baked apple that's been left over till it's all dried out. His hair wasn't braided like the Indians I'd seen in books, but hung down in scraggly strings to his shoulders, and he had a faded derby hat balanced square on the top of his head. His coat was faded black, too, and the back of it was long and rounded, like the minister's back in East Rochester. He had on a tight pair of bright blue pants, and white moccasins with lots of red beads on them. He just looked at me without changing his face a bit. Then he grunted and went over and sat down with his back against the side of the barn.

I had just got on my feet when Father came out to see why I hadn't brought in the milk. I must have had quite a bump on my head, because it was the first thing he saw, and asked me what kind of tricks I'd been up to now. I didn't get a chance to tell him, because the old man with the whiskers told him first. He said, "Me and Two Dog was a-sittin' here agin the barn havin' a smoke when this little coyote come a-ridin' in. The mare spooked and hightailed into the barn like a scairt prairie dog into his hole. The papoose, he didn't have a chanct and bunged his head agin the barn. Ain't hurt a mite, though; not a broke bone anywheres." He took hold of my arm and worked it up and down like a pump handle to show Father.

Father said for me to come to the house and get cleaned up for supper. On the way in he told me the old man was Mr. Thompson, and that he claimed to have had his camp site in 1840 right where our house was now. While I was getting washed I asked him about Two Dog, because he was the first Indian I had seen close to. Father said he didn't know much about him, but he had heard that he was a Blackfoot, and that he and Mr. Thompson had lived together up in the foothills since long before anybody could remember.

The house smelled awfully good. Father had killed one of the hens, and Mother had it cooking in the big iron pot. She was putting in the dumplings when she told us that Mr. Thompson and Two Dog were going to eat supper with us and stay all night. She said they might not eat just the way we did and she didn't want to catch one of us staring at them. Then she told Philip and me that she had fixed a shakedown in our room, and that they were going to sleep with us, and that she didn't want us to do any whispering after we went to bed because we might disturb our visitors.

Mother let me go out to call them when supper was ready, but Two Dog wouldn't come to the house. He was still sitting on the ground with his back against the barn. His eyes didn't move or blink, but looked off across our bean field as though he were watching something far away. Mr. Thompson said Two Dog wasn't used to houses and didn't like them, but he'd bring his supper out to him when we got done eating.

The way he ate, I don't think Mr. Thompson could have had a square meal in months. He just used his fork to push things on his knife, and he pushed them on clear up to the handle. Mr. Thompson kept telling Mother that he hadn't tasted such victuals since he was a little boy back in Missouri, and she kept asking him if he wouldn't have some more of everything. Every time she asked him he would pass his plate back and Father would scrape around in the nappy some more. I knew he was keeping one drumstick back for Two Dog, because he had only fished one out, and Philip got that.

When everything in sight was gone, Mr. Thompson tilted his chair back on its hind legs and wiped his whiskers with the corner of the tablecloth. Then he began telling us about the time he first made his camp right where we were sitting. I liked to hear him talk, but I was worried about Two Dog's supper, and asked Father if I could take it out to him. Father dished it up and Mother got me three biscuits and some mashed potatoes she'd been keeping hot on the stove.

I put the silverware and napkin in my overall pocket, so I'd have one hand for the plate and the other for the teacup. Two Dog hadn't moved an inch. He was still looking out across the bean field, but when I passed the plate out to him he looked up and his eyes smiled, but not his mouth. He took the plate with both hands and sat it down beside him, then reached them up for the cup. Instead of holding it by the handle, he took it like a bowl and tasted the tea. Then he looked up at me and said, "Shoog," but I didn't understand what he meant till he put one finger up above the cup and moved it around as if he were stirring.

I forgot all about his silverware and napkin, and ran to the house for the sugar bowl. When I got back he was still holding the cup like a squirrel holding an acorn, and looking across the bean field. There was about a cup of sugar in the bowl. He poured nearly a third of it into the tea and started to stir it with his fingers as soon as he had put the bowl down. Then I remembered the silverware and held it out to him. He looked at it a minute and then stirred the tea again with his finger. I didn't want to leave him and I didn't want to just stand there holding the napkin and silver, so I sat down beside him.

He finished all the tea first, then ate just the chicken leg off the plate. When it was gone, he took up the sugar bowl and poured a few grains into his cupped hand. He picked it out of his palm with his lips, like a horse picking the last few oats out of his feedbox. I don't know how long we sat there, but it was until long after the sun had gone down. Every ten or fifteen minutes Two Dog would pour a few more grains of sugar into his palm and pick it out with his lips. They were so dry they never made his hand sticky. He didn't say a word till the bowl was empty, and I didn't either. Once he put his hand over and let it drop on my knee; he lifted it slowly and let it drop twice more. When the bowl was empty he passed it to me and said, "Friend." That was all the conversation. I got up and went to the house with a lump in my throat and a big love in my heart for Two Dog.

When I came into the house, the supper dishes were done and all the children in bed, except Grace. Mother had the corn popper out and was popping corn over a hot fire. Mr. Thompson was still tilted back in his chair, but had swung it around and had his feet crossed on the window sill. He and Father were munching popcorn, but Grace was sitting with her eyes bugged out, and not even nibbling at the handful of popcorn she was holding. I sat down beside her and she leaned over close to my ear and whispered, "Oh, can he tell stories! He used to go hunting and fighting Indians with Kit Carson."

Mother had never let us sit up so late as we did that evening, and I had never seen anybody eat so much popcorn. Mr. Thompson seemed to have known every trapper and hunter who came west for beaver and buffalo skins. Between mouthfuls of popcorn, he told us about guiding wagon trains from Westport Landing to

Oregon, and about going to rendezvous on the Green River with Kit Carson and Bent and Lucien Maxwell. Every once in a while he would stop and tell Mother that he hadn't had such a fine evening since he was a little boy in Missouri. Then he would eat more popcorn, and start all over again.

His last story was about a fight with the Blackfoot Indians. He told how the Indians set fire to the prairie clear around their wagon camp, and about his being the only white man to get out alive. The only reason he didn't get killed was that Two Dog was a chief's son. Mr. Thompson pulled him out from under his dead horse just before the fire reached him, and two young braves rode in through the flames to save them. He said that was why he and Two Dog were blood brothers. I didn't know what that meant, so he told me how they had cut themselves and placed the wounds against each other, so their blood would mix and make them brothers forever.

After that Mother made Grace and me get ready for bed. But while we were brushing our teeth, Mr. Thompson kept on telling Father about Two Dog. He said, "Old Two Dog, he's the cleverest man with horses ever you see. That old Injun, he can take a horse critter that's nine parts dead, and have him prancin' 'round like a colt in a couple days. And there ain't no horse so mean he can't handle him." I wanted to stay and hear more, but Father snapped his fingers.

I couldn't have gone to sleep when I got out to the bunkhouse if I'd wanted to, and I didn't want to. I thought maybe Mr. Thompson would tell another story when he came out to go to bed, or that when he was putting his nightshirt on I might be able to see some of the places where he had been shot. He came out just a little while after I was in bed, but all he said was, "Whoosh." And all he took off was his calfskin vest and his high-heeled boots, then he crawled in between the blankets with everything else on. I asked him if he wanted me to go out and call Two Dog to come to bed, but he said, "Old Two Dog, he ain't never slept in no house; he'd rather sleep right where he's at." In two minutes he was snoring so loud I couldn't go to sleep.

Even if it was June, it was cold at night, and I got thinking about Two Dog sitting out there beside the barn. After a little while, when I knew Grace would be asleep, I pulled my overalls on and took the top blanket off our bed. I didn't make enough noise so I could even hear myself above Mr. Thompson's snoring, and was holding my breath as I eased the door open, but he sat bolt upright when I touched the latch, and said, "Who goes there?" I told him and said I was just taking a blanket out to Two Dog. He was snoring again before I closed the door.

When I got to the barn Two Dog was sitting exactly the way he was when I left him. The moon was way over in the west, above the mountains, and I could see that his eyes were open and that he was still looking off across the bean field. I held the blanket out toward him and he reached up for it. With a quick flip, he flung it around his shoulders so that it covered all but his head like a tent. He didn't say a word and I didn't want to just walk away and leave him, so I sat down beside him again. I guess I sat down by him because I was thinking about Mr.

Thompson's story, and wished I could be a blood brother to Two Dog as he was. He didn't look over at me, but he flipped the blanket around me so we were both under the same tent.

Grandfather used to be deaf and, before he died, he and I used to play sign language. I thought maybe I could talk to Two Dog in sign language, so I raised my eyebrows, put both palms together and laid my face down against them; then I looked far off along the mountains. Two Dog knew I was asking him where he slept—where his home was—just as well as Grandfather would have known. He pointed with a straight arm and finger toward the V-shaped gap where Turkey Creek came out of the mountains. Then, using his forefingers to follow the trail forward and upward, he told me where his camp was, high in the upland valley.

Two Dog and I sat and talked with our hands till the moon dipped down and started to slide away behind the mountains. Then he reached over and laid his hand against my leg in three slow strokes the way he did when I brought his supper. I knew he meant for me to go in to bed, so I went. But I stayed awake a long time, thinking about the stories he had told me with his hands. When Father called me in the morning, Mr. Thompson and Two Dog were gone.

Mrs. Corcoran came to the corral when I went to take the cows out the next morning. She hadn't combed her hair and had her hands rolled up in her apron because it was still a little chilly. She called, "Little boy!" before she was halfway across the dooryard, and from the way she had her mouth clamped up I thought she was going to scold me. But when she got over to where I was waiting to let the cows out of the gate, she said, "I hear tell that old reprobate, Horsethief Thompson, and his Injun put up at your place last night. Good lands! I hope your folks had better sense than to let 'em in the house. Did they?"

I told her that Two Dog wouldn't even come into the house for supper and that he slept sitting out beside the barn. She snorted like a spooky horse when I told her that, and said, "Then you're telling me your maw did let old Horsethief in? My land o' Goshen! Well, you better tell her they'll steal anything that ain't red-hot or nailed down. I hear tell that dirty old man's got more lice on him than a settin' hen. I suppose your folks didn't have no better sense than to sit around with their mouths a-gap listening to a pack of his lies."

I got so mad when she said that, that I forgot she was my boss, and hollered, "My folks have got more sense than you have, and he did not tell us a pack of lies. He told us about Kit Carson, and I know it isn't lies, because I read about Kit Carson in a book. And besides, he isn't a dirty old man."

I kept getting madder and madder at Mrs. Corcoran for what she said, until the lump got so big in my throat that I thought I was going to cry. So to keep from it, I slammed the gate open and ran Fanny right in among the cows. Until I had them nearly out to the wagon road, Mrs. Corcoran kept yelling after me, telling me that if she had a "sassy young one" like me she'd take him across her checkered apron, and saying more mean things about Mr. Thompson and Two Dog.

When I went past Aultland's for the milk that night, Fred told me he was going to start stacking alfalfa the next day and that Father was going to help him. Then he asked me how much Mrs. Corcoran was paying me for herding cows. When I told him, he said, "I'll double the ante if you want to ride stacker horse for me."

I didn't know what he meant, but I told him I'd do it, so he had me go back and tell Mrs. Corcoran I wouldn't be coming to herd her cows till haying was over. From her telling me she'd take me over her checkered apron if I was her young one, I didn't think she'd care if I never came back, but she just about had a fit. She asked me how much Fred was going to pay me, and I said I didn't know but he had said he'd double the ante. When I told her that she got madder than ever and called him a help stealer, and said he was ruining me so I wouldn't be any good to anybody. Then she told me she never wanted to lay eyes on me again. But when I was riding back to the road, she yelled for me to be sure and come back the day after Fred got done haying.

I didn't want to tell Mother what Mrs. Corcoran said about Mr. Thompson and Two Dog, because I knew if I did she'd get around to where I'd have to tell her that I'd been saucy. But I was worried that some of it might be true, and besides I didn't want to have any question in my own mind about either Two Dog or Mr. Thompson, so I told Father all about it while we were out feeding the horses.

Father said that, of course, you never could tell by the looks of a frog how far he'd jump, but he'd bet that neither Mr. Thompson nor Two Dog would ever steal anything from us, and that he thought Mr. Thompson was telling the truth in his story. Then he said Mother could get books in the Denver Library that would show whether or not Kit Carson did the things Mr. Thompson said he did and, if he did, then we would know the stories were true. He said maybe the part about the lice was right, but it might be best not to mention it to Mother till we knew more about it.

11

HAYING

I LIKED WORKING FOR FRED AULTLAND. HAYING AND threshing were big times at his place, and he always had a dozen or so men to help him. Some of them were neighbors who didn't have so much hay of their own, and some were hired hands Fred brought out from Denver. Father and I didn't work for him until the hay was all cut and raked into windrows. I had never seen a hay stacker before, and Father had to snap his fingers at me twice during the morning, because I got so interested in what was going on that I forgot about my own job. Fred's was what they called a bull-stacker and the hay was brought in from the fields with bull-rakes.

They were sort of three-wheeled carts, and always looked as though they were going backwards, because they scooped up the hay and carried it to the stacker in front of the horses, instead of behind them. Each load weighed nearly half a ton.

The stacker looked like the mast of a ship mounted on

a big turntable, with a long boom fastened near the bottom of it. The cradle was hinged to the end of the boom, and pulley ropes ran between it and the top of the mast. Jeff was the engine that furnished the lifting power, and I was the engineer. Jeff was a big, lazy old horse—strong as a pair of oxen—and had been pulling the hoist rope for the past five years. As Jeff pulled on the rope, the hay was raised from the bull-rakes and lifted nearly to the top of the mast. Then, while we held it there, Father and another man heaved the turntable around with a long gee-pole, till the cradle was over the stack. When I backed Jeff to slacken the hoist rope, the cradle tilted forward and the load fell with a thump. It took Fred and two other men to get it untangled and built into the stack before another load was brought in. The only hard part of Father's job was heaving the turntable around, but that made him cough a good deal.

After the first couple of loads, he talked to Fred about the stacker, and they sent a man to the barn for tools and other things Father needed. He worked, between loads, all morning; changing pulleys, rigging a heavy cable from the turntable to the hoist rope, and putting trip-catches on the cradle. When he was finished, they didn't need to heave the turntable around any more, nor lift the hay any higher than the top of the stack, and Father could drop the hay wherever Fred wanted it, by just jerking a trip-cord. In that way Fred only needed one man to help him on the stack, and Father could do all the work on the ground alone.

I liked noontimes best of any part of the haying. When it came twelve o'clock, Bessie would hammer on an old wagon tire hung near the kitchen door. The sound would roll out across the hayfields like the ringing of a big bell, and after it had stopped, the echo would come back from the hills as though they were full of far-off churches.

The minute the bell rang the drivers would stop their teams wherever they happened to be and unhook the horses. It was always a race to see who could get his team to the barn quickest, so as to get them unbridled and fed, and be first at the washstand. It was out by the windmill, and Bessie always had three blue enamel basins, half a dozen flour-sack towels, and a bar of homemade yellow soap waiting for us.

Aultlands had a big porch on the east side of their house, with a row of apple trees that shaded it. In haying and threshing time, Bessie set a long table out there, and that's where we ate our dinners. At home, Father always served everyone and said grace before we started to eat, but that wasn't the way they did it at Aultland's.

As soon as we were down at the table, Bessie would start bringing out big platters of meat and fried chicken, and potatoes and vegetables, and bowls of gravy, and plates of hot biscuits and corn muffins. As quick as she'd set a platter down, somebody would pick it up, help himself, and pass it on to the next man. They came so fast that I could hardly help myself from one before another one caught up to me. Some of the platters were still pretty heavy when they got to me, and I could just barely hold them with one hand while I forked some off with the other. At first the men wanted to hold them for me, but they saw I didn't like them to, and let me handle my own platters. Mrs. Aultland was a real good cook, and I used to eat until I couldn't hold another mouthful.

The most fun came after we were done eating. We had to take an hour for dinner because the horses needed that much time to eat and rest. So, as soon as the last piece of pie was eaten, the men would lie down on the grass under the apple trees. Father didn't smoke, but all the other men would get out their pipes or Bull Durham, and talk or tell stories while they were smoking. Jerry Alder was the best storyteller. Sometimes he told stories so quiet I could hardly hear them, and they didn't sound funny at all, but all the men would laugh till the fat ones had to hold on to their stomachs. Even Father laughed sometimes when I couldn't see anything funny.

It was one of those noons that I found out about pheasants. There were lots of them, and they were so tame they'd come almost up to the haystack. I wanted to do some of the talking after dinner as the men did. So one noon I told Fred that if I had a gun I could shoot some of those pheasants for us to eat, and then his mother wouldn't have to kill so many chickens. Everybody laughed at me, and Fred said, "If you're going to do any shooting in Colorado, shoot a man. You can always call it self-defense, but if you kill a pheasant you'll spend the rest of your life in the hoosegow."

Fred Aultland's haying lasted two weeks—Sundays and all. I remember the last day of that haying better than any of the others, because so many things happened. The last day of haying or harvest or threshing is always the day when the most things happen. Maybe it's because everybody is happy if you had good luck, and if you didn't everybody's glad it's over with.

There was a fight after dinner that noon. One of the young fellows Fred brought out from Denver said something about Bessie that Jerry Alder didn't like. She had her back to us and was picking up the dishes, and she was leaning over so far that her dress was real tight across her bottom. The Denver fellow was looking right at it, then he winked at Jerry before he said whatever it was. Anyway, it was an awful hard fight. The Denver fellow was the biggest man on the job, and Jerry was next biggest. The first sock Jerry hit him, Bessie ran into the house and all the men got up on their feet, but nobody tried to stop them.

The other two Denver fellows were nearest to where they were fighting. Fred and Carl Henry went over and stood by them, but they didn't say anything. The big Denver man didn't hit so often as Jerry did, but he hit a lot harder. He took a longer swing and once he hit Jerry under the ear and knocked him down. I thought he was going to kick him while he was down, but Fred stepped in quick, and he didn't. Jerry rolled over and got right up again, and from there on he fought just like a collie dog.

He used his feet just about as much as he did his fists, but he didn't do any kicking like the other fellow. He'd go in quick and hit, and be out again before the bigger fellow could hit back. And he went around that Denver man like a fly going around a lamp chimney. I guess the big fellow got kind of dizzy turning around and around, trying to catch up with Jerry, because he started looking pretty groggy. Then, all at once, Jerry flew in with both arms working like the Pitman rod on a mowing machine. He got his head right against the other fellow's wishbone, and hammered him in the stomach till he went down yawping for air like a mud cat when you toss him up on the creek bank.

After the fight, Fred took the three Denver fellows over to the bunkhouse and paid them off, but I don't think he ever said anything at all to Jerry for fighting. And as soon as he had washed the blood off his face and got his breath, you wouldn't have known Jerry had been in a fight—except that his lips were kind of swelled up. He came back from the washstand and started to tell stories almost before he had found a place to lie down under the apple tree.

With three hands short, it was late before we had the last load of hay on the stack, so Father and I stayed at Aultland's for supper. When we were through eating, Fred told us to come into the house with him. We sat down by the table in the dining room, and Fred got out his checkbook. I knew Father didn't know how much he was going to get, because I heard Mother ask him, and he just said, "I don't know. I think he's paying the men he got from Denver a dollar and a half a day, but they're quite a bit stouter than I am right now." I hadn't wanted to ask Father what Fred meant when he told me he'd double the ante, so I didn't know how much I was going to get either, but I hoped he meant he was going to give me fifty cents a day.

After Fred got the ink bottle and a pen, he sat down at the table with us and asked me if I wanted to have a separate check, or if he should make one check for Father and me together. I wanted it to be a big enough check that we could buy a cow, and I was proud to have my pay go in with Father's, so I said for him to just make one check. He looked up at Father, and said, "All right then, Charlie, that'll make it a round sum. I figure Spikes is worth twice what Liz Corcoran was giving him, and you've saved me the wages of two men. Will fifty dollars square the books?" I was so excited I didn't even hear what Father said, and he had to tap me on the arm before I remembered to say thank you.

Father was as anxious to get home and show Mother the check as I was. He walked so fast I had to trot part of the time to keep up with him. We hadn't gone very far before he noticed I was having to trot, and scrooched down so I could get on and ride pickaback. I had always liked to have Father lug me pickaback before—and we were far enough from Aultland's house so that I wasn't afraid anyone would see us—but for some reason I didn't want to be carried that night. It just didn't seem right to be carried home when we were taking the check I had helped earn. Father understood how I felt, and he walked slow enough so I didn't have to trot any more, and let me carry the check home to Mother in my overall pocket.

There wasn't nearly so much fun in giving it to her as I had thought, because when we got there our old white horse, Bill, was sick. He was breathing so hard you could hear him all over the yard, and was pounding his head on the barn floor. Father took one look at him and said, "Blackwater. I'm afraid he's done for." Then he sent me kiting back to Aultland's for a bottle of spirits of niter.

12

I Go After Two Dog

THEN NEXT MORNING I WAS UP AS SOON AS THE FIRST light peeped over Loretta Heights. Mrs. Corcoran had told me to come back to herd her cows right after haying, but I had a different idea in my head. Bill was still just barely alive and I was going to get Two Dog to come and save him. Before anybody else was up, I went out and sat beside our barn where we had sat the night he and Mr. Thompson stayed at our place.

From there I could get the best look at the mountains when the sun first struck them, and before it got high enough to light the land between them and me. Mother had a stereoscope that you could put pictures into and move them to make far-off places come right up close. The early sun did the same thing to the mountains. I could shut my eyes and see just how Two Dog's fingers had shown me the way to his camp, then open them and trace the trail up through Turkey Creek Canyon so it seemed almost as though I had actually been over it. I got up and swiped a quart of oats for Fanny, so she could have them all cleaned up before Father came out to give the horses their regular breakfasts. By half-past six I started off up the road on Fanny as if I were going to the Corcorans', but I had three cold biscuits hidden in the front of my blouse.

All spring Father had talked about our driving up to the mountains some Sunday, but for one reason or another we never did it. They looked as though they started just a little way beyond the hill in Fred Aultland's back pasture. Turkey Creek Canyon was quite a way south, and the most direct wagon road ran along the west end of our place, past the schoolhouse and Carl Henry's. But I knew Father would never let me go alone, and I didn't want anybody to see me, so I headed west past Aultland's wheat field, then cut southwest across country, straight for the V that marked the mouth of the canyon. I knew better than to run Fanny up hills, but I was so anxious to get to Two Dog's and the distance seemed so short, that I lay tight down against her neck and we went up over Fred's big hill like a jack rabbit in front of a coyote.

Looking from the top of that hill, I could see a series of others, rising one beyond the other toward the hogbacks that stood before the real mountains. Until then there hadn't been any doubt in my mind that I could get to Two Dog's camp without a mite of trouble. But, with all those hills between me and the mountains, I began to get a little bit afraid, and wondered if I shouldn't go back and talk to Father about it first. As soon as we were out of sight over the top of the hill, I stopped Fanny and let her catch her wind. The more I thought about talking to Father, the more sure I was that he wouldn't let me go. And I was just as sure that Two Dog was the only one in the world who could save Bill, so I kicked my heels against Fanny's ribs.

At first there were crops in the valleys between the hills, and a few ranch houses, so I had to ride miles out of my way to get around them. Every time we got to the top of one hill, there was another just beyond it, and the mountains didn't seem any nearer than they had from home. I knew Fanny was beginning to get tired, because the hills were getting steeper and she was climbing slower. There were no more crop fields in the valleys, and I started riding around the hills instead of over them, so as to save Fanny the hard climbs. Two or three times we came to deep gulches that we couldn't get across, and had to turn back and find another way. If it hadn't been for the mountains I'm sure I would have been lost, but I knew their shapes well enough so that I could always tell where I was. It was getting close to noon and the sun was bearing down like a hot stove lid when we came into a green little valley with a spring of cool water in it. We both drank all we could hold, and while Fanny grazed I ate my biscuits. I must have squeezed them a bit, because they were pretty well crumbled up, and some of the pieces were soggy with sweat, but I was hungry and they tasted all right.

The sun was hanging low above the mountains when we came over the last hill and I could see the break in the hogback where Turkey Creek had cut its gorge. As we came closer I could see there was a little-used wagon road along the north bank of the creek. I loped Fanny toward it and we followed it through the gorge and into the mouth of the canyon. The misgivings I had when we were on top of Fred Aultland's hill were nothing to what I had when we came into the canyon. The creek ran through a narrow cut, and the walls seemed to rise straight up for a mile. From there, the sun had set and a cool breeze was drawing down between the cliffs. All I had on was my blue shirt and overalls, and after the heat among the hills, it made me shiver. I don't know whether I shivered more because I was cold or because I was frightened. I had never seen mountains that were more than big rolling hills, and it seemed to me that those black rock walls might fall on me any minute.

Then I really began to be afraid I could never find Two Dog's camp. I stopped Fanny and shut my eyes tight, trying to bring back the way he had pointed out the trail with his fingers, but all I could see was a big green blotch with black rock walls running up around it. As I had sat beside the barn with Two Dog a couple of weeks ago, and again that same morning, I had been able to picture the trail just as I was sure it was going to look, but it was all different now. For a minute or two I was going to turn back, but I knew night would come long before I could make it, and I could never hope to find my way home in the dark. I kicked my heels into Fanny's ribs and we went on. The harder I tried to think how Two Dog's fingers had moved, the more confused I got.

In half an hour it had become darker and colder in the canyon. I could remember that Two Dog's fingers had shown the trail going in quite a way before it branched off, but he had made them go straight, while the trail wound in and out against the wall of the canyon. At last I thought that if I could just be sitting down behind our barn again for a few minutes I could remember it all right. But, of course, I couldn't do that, so I slid off Fanny and sat down with my back against the canyon wall. I was so tired I almost went to sleep, and it must have been when I was just between being asleep and awake that it all came back to me. I could remember that he had shown the trail going up, as though there was a steep hill, and then angling off to the right. I climbed back on Fanny and put her into a good stiff lope. It wasn't more than ten minutes before we came around a shoulder of rock and the track climbed steeply up a shelf on the canyon wall.

Just above the rise the trail forked. The main track followed the shelf above the creek, but a thin thread of it turned up the side of a jagged cleft through the rocks to the right. I had no question in mind, and turned Fanny up the steep side trail. The sun had sunk so low that it no longer shone on the top of the peaks above me, and I began to get panicky for fear black darkness would catch me and we would fall to the bottom of the gorge if Fanny made a misstep. I dug my knees into her withers and kept her climbing so hard that it made her breath whistle through her nose.

We were nearly at the top of the climb when the whole air of the canyon was ripped to pieces by a sound that almost made my heart stop. It was a howl that seemed to come from nowhere in particular, but from everywhere at once, as it echoed back and forth between the canyon walls. Cold shivers raced up and down my back and it felt as though it were covered with stiff hair that was standing up as it does on a frightened dog. Fanny must have felt just the same way I did, because her ears pinned back tight against her head, and I could feel a tremble pass through her withers. She crowded close against the cliff and stood shaking.

I started thinking about Father and Mother and the rest

of the youngsters at home. I wanted to turn Fanny and race out of the canyon as fast as she could go, but when I looked down into the gorge it was as black as a well. Though I had never heard a wolf's howl before, I had read about it and knew that was what I must have heard. I tried to remember the sound and see if I could figure out whether it came from above or below, but I was so scared I couldn't think straight, and when I shut my eyes I could see gray shadows racing up the trail behind me. That settled it. I kicked my heels into Fanny's ribs and tried to cluck to her, but my mouth was so dry that I only made a hissing sound.

I think that was the first time Fanny ever trusted my judgment more than her own. She gathered her muscles and tore up the rest of the grade as though the wolf had her by the tail. We came out onto a flat rock ledge, raced across it, and were out onto a narrow path that wound through great boulders. Fanny was taking the sharp turns of the path so fast that I had to hang on with every ounce of strength in my knees. We must have gone a mile or more that way—I could hear every breath she took rasp through her throat like tearing cloth. It was deep twilight when we came out into a little open field set in between tall black-looking trees—and the path was gone. I sawed on the reins and pulled Fanny to a stop in the middle of the field. We stood shivering as though it were below zero. There wasn't a sound except the rushing of Fanny's breath. The first thought that came into my head was—*timber wolves*. I had read stories about their tearing wood choppers to pieces, and turned Fanny to get back out of there the way we had come in. But I couldn't even see a gap in the wall of black trees, and I was so panicky I couldn't remember whether there should be more to the trail or not.

Without even thinking what I was doing, I yelled, *"Two Dog!"* at the top of my lungs. The sound came yodelly like a coyote call. A second later an oblong of light from an open doorway showed at the edge of the woods, and Mr. Thompson's voice called out, "Hi there, Little Papoose."

Mother used to sing a song about "the golden gates of heaven," and that's what the yellow light coming out of the doorway reminded me of. I leaned forward a little bit on Fanny and she went over there on the fly. I guess that light looked as good to her as it did to me.

When I rode up to the door, Mr. Thompson told me to light down and come in while he put Fanny in the corral. At first I didn't want to let him take her and asked if the wolves might not get her, but he just laughed and said, "Ain't saw a wolf 'round these parts in years—'ceptin' Two Dog's old tame one. Always hollers when there's anybody on the trail, and generally scares 'em off. That's how we knowed you was comin'."

Their house only had one small room and not a single window. It was made of poles on the front and sides, and built right against a ledge, so that the back wall was solid stone. The spaces between the poles were stuffed with hard-baked adobe and straw. There wasn't any stove or chimney, but there was a cleft in the ledge about three feet deep that they used for a fireplace. It was wedge-shaped, and about as wide as it was deep at the bottom, but the top narrowed to less than a foot. The floor was partly a flat rock and the rest hard dirt. There were two bunks at one end of the room—one above the other—but there weren't any bedclothes or mattresses. The springs were made of tightly stretched horsehide, and the covers were mountain-goat skins with long white hair.

The only furniture was a table and two stools. The table must have weighed a ton. It was nearly four feet wide, and had been made by splitting the butt of a log in two. The legs were heavy stakes driven into holes in the round side of the log. One stool sat on each side of the table. They were made the same way, and didn't look as though they had ever been moved. Pieces of wagon iron, worn horseshoes, and harness hung on wooden pegs in the walls. Strips of dried meat and bunches of herbs were tied to a line in front of the fireplace. The only lamp was a bottle of fat with a rope wick in it. It didn't have any chimney.

Two Dog was sitting on the floor beside the fireplace with his back against the stone wall. He didn't get up when I came in, but his eyes lighted and he held one arm toward me with the palm of his hand down. I didn't know how to shake hands with his palm turned down like that, so I just took hold of the ends of his fingers, then let go and sat down beside him. He didn't say a word, but reached over and laid his hand on my leg three times, the way he did beside our barn. It was five or ten minutes before Mr. Thompson came back from putting Fanny in the corral. I had plenty of time to show Two Dog how Bill was lying on our barn floor with his back all humped up. And how he was pounding his head, and how he was breathing.

I used to wonder if the reason Two Dog didn't talk wasn't because Mr. Thompson talked enough for both of them. As soon as he came back from putting Fanny in the corral, Two Dog said about six words to him—kind of grunts, I guess it was Indian. Then Mr. Thompson began asking me questions faster than I could answer them. He wanted to know if Father and Mother knew I was coming up there, and how I had found their place, and if my folks wouldn't be worried about me. All the time he was talking he kept fussing with something in a big black iron pot over the fire.

While I was telling him, he took three dented old pie tins from the table and started ladling out the stew. It looked like rabbit stew, but the gravy was thick and brown. There was a covered iron kettle sitting on the floor by the fireplace. Mr. Thompson fished a few cold biscuits and three iron spoons out of it, put a biscuit and a spoon on each plate, and gave one to Two Dog and one to me. Then he sat down on a stool with his plate beside him on the table.

Mr. Thompson kept asking questions all the time between mouthfuls, and telling me to hurry and eat my victuals so he could take me right home. I was real hungry and the stew was good, so I just let him talk till I had cleaned up my plate. Just as I was sopping up the last of the gravy with my biscuit, Two Dog patted me on the leg again, nodded his head toward his plate, and said, "Skunk—good!" For a minute I thought I was going to be sick, but I decided it wouldn't hurt me if it didn't hurt them, and it stayed down all right.

As soon as Mr. Thompson was through eating, he

snatched up the stew pot and took it outdoors. I heard him clapping his hands before he came back. Two Dog got up, took a coil of thin shiny rope from a peg in the wall, and motioned for me to follow him out the door. As I did, my heart jumped into my throat and nearly stopped. A big gray wolf was eating from the iron pot. He was standing in the light that spilled through the doorway, and when he lifted his head his eyes glowed like live coals. He snarled, and the hair bristled on his shoulders, but Two Dog grunted at him and he faded away into the shadow of the trees.

The moon had risen, and Two Dog led the way along the woods at the edge of the field to a pole corral at its far end. There was a break in the trees, so that moonlight flooded the corral, and I could see nearly a dozen mean-looking horses inside it. Fanny and the two buckskins I had seen at our place were among them. They started milling when they saw us, and crowded into the far end of the corral, snorting and rearing against the poles. Two Dog motioned me to stay outside, while he crawled through the bars.

He seemed so frail and old that I was sure they would kill him, but he walked straight toward them. As he went, he shook out a loop in the horsehair rope, holding it in one hand and letting it trail along behind him. He was almost to them when one of the horses whistled, and they all came racing toward him. I ducked my head without meaning to, and when I lifted it Two Dog was snubbing one of the buckskins to a corral post. The buckskin jumped and reared, fighting the rope for a couple of minutes, but it didn't seem to worry Two Dog a bit. He waited for the bronc to calm down, then led him to the gate and haltered him.

I watched like a hawk when he caught the other buckskin, but I couldn't see how he did it. He didn't any more than snap his wrist and forearm, but the rope leaped off the ground, passed over another horse's back, and came looping down around the buckskin's neck. It all happened in less than a second. After that, he caught Fanny the same way, only he didn't have to snub her to a post. As soon as she felt the rope around her neck she stopped dead still. Two Dog snapped his wrist again and a loop that looked like a little barrel hoop ran up the rope and settled around her nose. Then he led her to the gate and put her bridle on.

I started to climb up the poles to get on her, but Two Dog shook his head at me. There was a rawhide strap about an inch wide hanging on one of the corral poles. He cut a piece off it a little more than a foot long, sliced about half its length into three narrow strips, and braided them into Fanny's mane, way back close to her withers. Then he showed me how to grab it with one hand and swing myself up so I could get an arm over her back. From there it was easy to pull myself on, and Fanny wore the rawhide braided in her mane as long as she lived.

Two Dog led the horses to the house, and when Mr. Thompson came out with the harness he was all dressed up in his calfskin vest, ten-gallon hat, and high-heeled boots. While he harnessed the horses, Two Dog went in and put on his black coat and derby. When he came back he was holding a small leather pouch that rattled as if it had dry leaves in it.

I don't remember much about the trip home that night. One minute I was listening to the drumbeat of the buckskin's running hoofs, and what seemed to be the next, Mr. Thompson was passing me over the wagon wheel into Mother's arms, and she was crying. I was awfully sleepy and I just remember having my head against her neck and telling her I was sorry, as she was carrying me through the bunkhouse door.

It was pretty late when Father came and woke me. He sat on the edge of my bed and held me on his lap. Then he told me how wrong a thing I had done, and that it had frightened Mother so that he wouldn't be surprised if it took several years off her life. He said that every man in the neighborhood had been out riding the hills looking for me, and that he thought Mother would have lost her mind if he hadn't made her believe Fanny would have come home alone if anything had happened to me. Then he said that wasn't really so, because she might have broken a leg in a gopher hole and fallen on me. I don't remember Father ever kissing me any other time, but after he put me back in bed he leaned over and kissed me right on the forehead. I didn't wake up till late the next morning. When I did, Mr. Thompson and Two Dog were gone, and Bill was up on his feet, nibbling at a few wisps of alfalfa.

13

WE GO TO AN AUCTION

THE SECOND SATURDAY AFTER I WENT UP TO TWO DOG'S there was an auction at one of the ranches down in Bear Creek Valley. Father and Mother were going to go and see if they could buy a good milch cow. While we were eating supper Friday night they were talking about the auction. Mother wanted to buy some things for the house, and some more chickens, and she said she'd like to get a turkey hen if she could pick one up reasonably. Afterwards they talked about what kind of a cow they'd like. Mother said to be sure it was one with a heifer calf, and she hoped it would be a Jersey because they gave good rich milk. Then Father said, "Mame, don't you think Ralph ought to go along and help pick out the cow, since he's earned part of the money?" I was drinking milk when he said it and caught my breath so quick I pretty near choked to death. As soon as supper was over I rode up to the Corcorans' on the fly to tell them I wouldn't be there to herd the cows next day.

Mr. Wright was the auctioneer, and they started off by selling furniture and pots and pans and things out of the house. Mother stayed there to see what bargains she could find, but Father and I went to the corrals and barns to look at the stock. I guess everybody who lived within ten miles was there. Mrs. Corcoran was in the corral looking every cow over and feeling their bags, and Fred Aultland and Jerry Alder were in the barn looking at the horses.

Fred came out and stood beside us while we were looking at the cows. There must have been thirty of them. He put his foot up on the bottom rail of the fence, laid his arms on the top one, and rested his chin against it. "Charlie, there are two or three pretty good cows there," he said. "Why don't you buy that brindle over there that Liz Corcoran's looking at, and this big Holstein nearest us? They'd give you all the milk and butter the kids could handle."

Father didn't say anything for a minute, then he smiled at Fred, "Did you ever hear of the fellow who could have bought Brooklyn Bridge for a million dollars, only he was short nine hundred and ninety-nine thousand, nine hundred and ninety-nine dollars and fifty cents?"

Fred laughed and said, "You can swing it all right, Charlie. I don't think those two are going very high, anyway." Then he rumpled up my hair, and said, "I'm going to have a couple more hayings this year, and you got Spikes here to help you pay for them. I hear tell he's holdin' Liz up for thirty-five cents a day now." He said it loud enough so that she couldn't help hearing him, and then he laughed till you could hear him all over the place.

Father just chuckled a little. Then he said, "It isn't only the cost of two cows, Fred. I'm going to have to ration close if I have enough winter feed for three horses and one cow off that little patch of oats and alfalfa."

Mrs. Corcoran got red as a beet when Fred laughed about me charging her thirty-five cents a day for herding her cows, but instead of going away she started looking at cows nearer to us, and I think it was just so she could

120

listen. Anyway, Fred let his voice way down and said, "I can stop you on that one, too. Soon as cold weather sets in I'll go to baling all the alfalfa I'm not going to feed to my own horses. There's always a lot of leaf and chaff shakes out when you bale alfalfa. It's too dusty for horses, but it's great cow feed. Liz'll only give me a dollar and a half a ton for it, and I'd rather sell it to you. Those cows would turn it into some damn cheap butter."

It was the hay part that convinced Father. He left me to look at horses and hogs with Fred and Jerry while he went back to the house to talk with Mother.

I don't know when I ever ate anything as good as we had at the auction. They had dug a big pit out by the windmill and built a fire of railroad ties in it the night before. And there was a whole yearling calf roasting over the pitful of red-hot coals. They had a windlass rigged up over the pit, and the whole dressed calf was trussed on a turn-bar. An old man with a big walrus mustache was turning the windlass and throwing handfuls of salt on the meat as it turned brown. You could smell it all over the yard and it made me almost drool.

It was about noon when Mr. Wright finished auctioning off all the things from the house, and the hens and ducks and turkeys. Then two or three men helped lift the roasted calf from over the pit and put in on a big heavy table. There were boxes and boxes of soft round rolls and two or three firkins of butter set out on another table. They brought out four or five butcher knives and put them on the table with the meat. Then they brought a washboiler full of coffee from the house, and pitchers of milk, and pies, cakes, and doughnuts. Everybody who could get near enough to the meat grabbed a knife and sliced off big wedges to put in the split rolls and make sandwiches. I was one of the first ones to get a sandwich. Jerry Alder put me on his shoulder and went through the crowd around the table like a Shorthorn bull going through a pack of coyotes. I ate so much my stomach ached clear up to my wishbone.

As soon as dinner was over Mr. Wright started the horse auction. There were some pretty good horses, and Fred paid more than a hundred dollars for one of them. He was a three-year-old bay, and almost as big as old Jeff that I rode to pull the stacker. Everybody stood out in a circle in the yard, and they led the horses out one at a time. Before anybody bid on it, the man who was having the auction came out into the circle and told about the horse. He would tell how old it was and how long he had owned it and how well it would pull and all kinds of things. If it was one he had raised from a colt, he would tell which mare was the dam and what stallion the colt was after. Every time he finished telling about a horse he'd pat it with his hand and say, "Sound as a nut!"

We couldn't be buying any horses, so we stood at the back of the circle. I guess the horses were all bringing more money than Father and Mother had thought they would, because I could see them keep looking at each other every time the bidding went above fifty dollars. After a while Mother whispered, "With horses selling at this kind of price, the cows will probably be outrageous. I do hope there will be one cheap enough so that we can afford her, but we'll have to have money for groceries this winter."

I was standing on Father's side and couldn't hear what he said when he turned his head to whisper to her, but she didn't watch the horse auction any more; she just kept looking down at the ground. In a little while she leaned over close to him and said, "I'm afraid I spent more than I should have on things for the house, but there were some lovely bargains." Then Father sent me to look at Fred's new horse while he and Mother took a little walk.

I had drunk so much milk with my sandwiches and doughnuts that I had to go awful bad, so Fred took me around by the cow corral while they were auctioning off the last horse. While I was busy he went over and talked to the man who had been leading the horses out to the ring. They were looking at the cows when I got back, and I saw Fred slip a silver dollar into the man's hand when he turned away from the corral.

Father and Mother had come back, and we stood right in the front row for the cow auction. Father stood next to Mother and then came me, and then Jerry and Fred and Mrs. Corcoran—Mr. Corcoran had to stay home and herd the cows. They brought out one cow after another, and they sold for anywhere from thirty to fifty dollars apiece. One or two old skinny ones sold for a dollar or two under thirty, but the man who was having the auction bragged about every one of them.

I kept watching for the two cows Fred had pointed out to Father, but I didn't see them. I wasn't too sure I could recognize the Holstein, because there were ten or twelve black-and-white cows sold, but there was only the one brindle in the corral, so I knew I wouldn't miss her. Mrs.

Corcoran bid on almost every cow, but she always stopped when the bidding got up to thirty or thirty-five dollars. It seemed as though they must have brought out all the best cows first, or else everybody got cows who wanted to pay lots of money for them, because when it was getting along toward the last, nobody was bidding much over thirty dollars. I thought sure they'd come to the end before the man Fred had been talking to led out the big Holstein. I knew her the minute he led her into the ring, and poked Father on the leg.

Mrs. Corcoran stepped right forward a foot when the man led the Holstein in, and she bid twenty dollars for her the first crack out of the box. Father said, "Twenty-one," somebody else said, "Twenty-one fifty," and then Mrs. Corcoran yelled out, "Twenty-five dollars." I knew she was going to bid more than we could pay, and hung my head down. I think I was saying a little prayer that she'd stop at thirty, when I saw Fred Aultland step right on her foot. She jumped and glared around at him, but she didn't bid on the Holstein cow any more. Father said, "Twenty-five fifty," and somebody said, "Twenty-six," but we got her for twenty-six fifty.

Before they led out the brindle cow, Mrs. Corcoran had moved down the circle away from Fred, and she yelled, "Twenty-five dollars," before the man was through telling how good a milker the cow was. I was watching her and didn't notice Fred moving down there, too, till he came right up beside her. I guess I wasn't the only one who saw him step on her foot that time, because some young fellow on the other side of the ring called out, "Get her again, Fred!" Everybody started to laugh and Mr. Wright yelled, "Sold," just as soon as Father said, "Twenty-six dollars."

I helped Father harness Bill and Nig back to the wagon while Fred and Bessie Aultland helped Mother collect the things she had bought. Father let me sit on the back of the wagon and lead our new cows home. He seemed happy while we were loading the things into the wagon, but Mother didn't say much. She had her lips buttoned up tight and her face was bright red. On the way home she talked most of the time, though. I couldn't hear all she said, because the Holstein cow held back on the rope, but I did hear enough to know she had spent more than she thought she should have. She said, "I just couldn't let those lovely Buff Orpington pullets go by at twenty-five cents apiece." And, "Two dollars and fifty cents does seem a lot of money to spend for two turkey hens, but Bessie says they're good foragers and will cost hardly a penny to feed. If I have good luck and am able to raise a brood of young turkeys, they should furnish us some very inexpensive meat—and it's so nice to have a turkey for Thanksgiving."

Then she said something about it probably not being necessary to spend the two dollars for a chest of drawers for the girls' room, but it was solid walnut. The first thing I heard Father say was, "That's a nice-looking little cuckoo clock you got." I looked around when Father said that, and saw the red run right up Mother's neck. Then they both laughed, and Mother said, "Don't you josh me about that clock, Charlie. I know we didn't need it, but it looked so much like home, and I just got bidding for it against Mrs. Thied and some other lady, and couldn't stop."

Grace had seen us coming when we were half a mile away, and all the youngsters came running up the road to meet us. She and Muriel had Hal by each hand and were almost dragging him along. I guess we all felt we were kind of rich people to be able to buy all those things. Philip put his bid in right away to be allowed to herd our cows.

14

THE IRRIGATION FIGHT

THE IRRIGATION FIGHT BROKE OUT SOON AFTER WE GOT our cows. July was hot, the creek was low, and there was only half a head of water coming through the ditch. It started at the gorge where Bear Creek came out of the mountains, and each ranch, all the way down, had rights to so many running square inches of water. Some of it soaked into the ground as it moved along, and some was drawn up by the sun, but—unless the creek was very low—there was enough for everyone to take his full measure. Each ranch had its own ditch box. They were wooden chutes that the whole body of water passed through. And each chute had a spillway with a gate to let out the full measure of that rancher's water right. There were gauge marks on the boxes and, when the ditch was running less than full, each man was supposed to set his gate so that it would take only his share.

There were water hogs near the head of the ditch. They were men who would take their full measure of water, and more, too, when the creek was low and crops were burning up. There had been a feud between the ranchers at the head and tail of the ditch ever since it was built. The first that I knew much about it was one night when Fred Aultland came down to talk to Father. They had a deal where Fred used all the water for both ranches twenty days, and then we had it for one. Fred came the evening before our day to have the water. Our oat field was so dry that Father was afraid the kernels wouldn't fill unless the ground got a good soaking right away. And the leaves on our peas and beans were curled up and withering. The vegetables from Mother's garden were little and scrawny.

I knew there was something the matter when I saw Fred coming down the wagon road. He always drove his tall bays as fast as they could trot, but that night they were just moping along. A cloud of dust was rising from the wheels of the buckboard that looked like white smoke from a bonfire. And Fred was hunched over with his elbows resting on his knees. Father and I went out to meet him as he came into the yard, and he looked terrible. One of his eyes was swollen and black, and there was dried blood around his mouth and nose. I started to ask him what the matter was, but Father laid his hand on my shoulder, so we just waited for Fred to talk first.

After a minute or two, he looked up at Father, and said, "Charlie, I'm afraid I've started something I'm not big enough to finish. For the last three days there hasn't been more than a trickle of water coming through the ditch as far as my place. Jerry Alder, Old Man Wright, and I went up this morning to have a look. Hardesty and Hawkins, both, had their sluice gates wide open and were taking double their full measure. Kuhl had his gate wide open and had made a cut in the bank where the ditch is built up at the back of his alfalfa field. He tried to tell me it was a natural break and he didn't know it was there, but there were shovel marks in the bank. The cut's filled in now—or it was when I left there."

Fred didn't say anything for a minute or two, and Father said, "Isn't there any court you can appeal to?"

Fred squirted a line of tobacco juice down between the off horse's heels and kept looking at the place where he had spit. After a while, he said, "Yep. We could haul 'em into court, and every one of 'em would lie like hell and say they never took more than the water that goes with their land. It would be their word against ours and we wouldn't have a snowball's chance."

Father said, "Can't you take enough witnesses up there to see what they're doing and outweigh their testimony in court?"

Fred spit between the other horse's heels, and said, "Yep, but they've all got their land posted, and we couldn't see what they're doing without getting on their land. They'd get us as bad for trespass as we could get them for stealing water."

Father said, "Isn't there a way of proving how much water passes through the ditch at their upper boundary and how much they let pass beyond them?"

Fred seemed to be thinking about it for a minute, then he looked up at Father, and said, "Look, Charlie, this is the hell of it for you and me: The water goes with the land. Your deed says you are entitled to so many inches of water, or *'such proportion of same as may be available!'* There's nothing that says whether that means available at the head of the ditch, or available at your own sluice gate. It's been fought ever since there were courts and ditches here, and there are rulings both ways. Every damned one of us would break himself if we tried to fight it clear through the courts. And, besides, if we'd get a high court ruling that it meant available at our own sluice boxes, these ranches at the end of the ditch wouldn't be worth a damn an acre. There's only one way to do it; we've got to take the law into our own hands and fight the water through the ditch. And, by God, I'm going to start tonight."

Father didn't like me to be around when men were swearing, and Fred looked mad enough to begin any minute; but before he did, Father sent me to get our cows. They were picketed out on the prairie near the railroad track.

During supper Father hardly said a word. Mother didn't eat much and kept biting her lip the way she always did when she was nervous. Father always milked the Holstein and I milked the brindle. While we were milking that night, I asked him if he was going to do anything about the water. He didn't answer me for a while, then he said, "Son, there are times a man has to do things he doesn't like to, in order to protect his family." He didn't say any more, and I didn't think I ought to ask him.

Something woke me in the night. It must have been after moonset, because it was dark as pitch. I lay listening for a long time, then I thought I heard a man's voice from over toward the railroad track, so I got up and looked out the windows. There were three little lights moving around in our oat field, and two more in our bean field. They were so far away that they looked like fireflies. I thought Father ought to know about it, and went into the house to tell him. Mother was wide awake, but Father wasn't in bed with her. She told me to go right back to bed because I needed my rest, and that the lights out in our fields were all right.

I went back to the bunkhouse, but I didn't go back to bed. I pulled my overalls on over my nightgown and tip-toed out into the darkness. I knew Father was out there somewhere with a lantern, and I wanted to see what was going on. When I was almost to our oat field, the lights all came together in one place and moved up the railroad toward Fred Aultland's. I thought I heard water gurgling among the oats, and when I went a little closer my foot sank ankle-deep in soft mud. While I was standing there watching the lights from the lanterns grow smaller and smaller up the track, I heard the sound of half a dozen rifle shots from way off toward the west. I was worried about Father, and afraid he might not be one of the men with a lantern, but be farther up the ditch where there was shooting. I wanted to run after him and tell him to come home, but I was scared and went back to the bunkhouse.

I didn't sleep another wink all night, and when it was just light enough so that I could see the outline of Loretta Heights against the eastern sky, I heard Father come home and go in the kitchen door. I couldn't see him and would never have known him by his walk. His feet sounded as though they were dragging, and he had on rubber boots. I heard him take them off before he went into the house. A little later I heard him coughing. It was that dry, hollow cough he had after the windstorm. As soon as it was light enough to see in good shape, I got up and got the milk buckets. I milked both cows, watered the pigs, and fed the horses before I went in to breakfast. I got cramps in my hands before I got done milking the big Holstein. She gave a bucket brimful of milk, and her teats were large with little bits of holes in them.

Father wasn't up when I went to work, and at breakfast Mother wouldn't talk. She kept biting her lip and her eyes looked as though she had been crying. I didn't see a soul around Corcoran's place when I let the cows out of the corral, and there was nobody in sight when I went past Aultland's. The road was all muddy where our ditch went under it. The culvert was a good big one, too, so I knew there must have been a terrible head of water come down through there during the night.

I didn't see a moving thing, except the cows and Fanny, until Grace brought out my dinner pail. She said Father had just got up and that there was a big red lump on his forehead, and he had been coughing in his sleep all morning. Grace could usually get Mother to talk, but she hadn't been able to find out a thing. Mother had made her play out in the back yard with the other youngsters all morning. I told her about the water in our oat field, and the lanterns and the shooting up beyond us on the ditch. I thought maybe the lump on Father's forehead was where he had been hit with a bullet, but Grace said it wasn't. She had read lots of books about wars—she liked them best of all—and she said she'd bet it was where he had been hit with a clubbed rifle.

There must have been some terrible battles up the ditch those next few nights. Father would leave the house just after I went to bed, and wouldn't get home till nearly daylight. He had another big lump on his cheek-bone that turned black and blue, and Fred Aultland and Jerry Alder and Carl Henry looked all beat up when I saw them. Jerry had his right arm in a sling.

Saturday night there was a meeting at our house. Men came from all the ranches west of us—halfway to the mountains. They must have started getting there just after I went to sleep, but I woke up when the first buggy came into our yard. It was Mr. Wright. I knew his voice when Father went to help him unhitch his team, and I knew there was going to be some kind of meeting, because the first thing Mr. Wright said was, "Ain't any of the rest of the fellas got here yet, Charlie?" And besides that, Mother had put Hal out to sleep with Philip and me.

Grace didn't get up till the third team came. Then she tiptoed into my room and we peeked out under the cur-tain together. The men all stood around the barn and talked till Carl Henry came—he was the last one—then they went into the house. Grace and I knew we shouldn't have done it, and that we'd get a good spanking if we got caught, but we crept out the bunkhouse door and crawled around to the kitchen window. It was open, so we had to hunker up against the side of the house and keep real quiet.

At first, everybody was trying to talk at once, and

someone said the only way they could ever keep water coming down the ditch in dry spells was to put men with high-powered rifles up on the hills, and shoot hell out of any so-and-so that went tampering with a ditch box. Then somebody else said that wouldn't do any good because the sheriff would get out a posse and throw them all in the hoosegow. They talked, and talked, and talked. Some of them even shouted, but I didn't hear Father's voice till Fred Aultland said, "Charlie, you must have done some thinking about this, but I haven't heard you say anything."

Everybody got real still then, and Father talked so low we couldn't much more than hear him. He said, "Well, it seems to me that courts are usually the best places to settle disputes if men can't get together peaceably, but in this instance both sides are afraid of what the court's ruling might be. We've been able to fight enough water down through the ditch at night to save our crops for the moment, but that won't do in the long run, because, sooner or later, somebody's going to be killed. When that happens, the matter will be settled in court whether we like it or not. It would be my idea that we ought to sit down and try to work out our differences with the men we've been fighting."

The men didn't seem to like that at all, and started shouting and talking all at once again. Some of them even swore—with Mother right in the other room. Mr. Corcoran called the men up near the head of the ditch some awful names, and said you might as well argue with a jackass as any one of them. At last Mr. Wright had to pound on the table and shout, "For God's sake, shut up and give Charlie a chance to tell us what his idea is, anyway."

Father didn't start to talk again till everybody was quiet, then he said, "Those fellows up there are holding the trump cards and they know it. I'm not too sure I wouldn't take pretty near my full measure of water if I were in their places and saw my crops drying up. I don't think they want a court fight, or a fist fight, or a gun fight any more than we do, but I don't think they're going to give up the hand without winning the odd trick. I wouldn't do it, and I don't think any of you fellows would. I'm inclined to think we'd be better off to have the assurance of a reasonable part of our share in dry time, than to take the chance of not getting any and losing all our late crops."

Father stopped talking as if he expected them to say he was wrong, but nobody spoke till Mr. Wright said, "Go on."

Then Father said, "I believe that if we approached them right with an agreement that we'd settle for 80 per cent of our proportion, based on ditch-head level, we might come to terms with them."

Jerry Alder and two or three of the younger fellows thought it would be better to keep on fighting the water down the ditch at night, but Mr. Wright, and Fred, and Carl, and even Mr. Corcoran thought Father's idea was best. It was right then that Mother pushed up the window in the front room, and Grace and I got scared, so we had to crawl back to the bunkhouse. In about half an hour all the men came out and started hitching up their horses. Mr. Wright was the last one to drive away, and before he went, he called to Father, "You'll be at my house, then, at ten o'clock tomorrow morning?"

Father called back, "I'll be there," and went into the house and closed the door.

There weren't any more fights over water that year, and when Willie Aldivote came up to the pasture to visit me a few days later, he seemed to think Father was quite a hero. I was proud because he said Father could fight like hell for a sick man, and that everybody thought he did a smart job getting the men up the ditch to agree about the water.

15

I Give Mr. Lake a Ride

ABOUT THE ONLY FUN I HAD THE REST OF THAT SUMMER was the two times Fred Aultland put up his hay. Father and I worked for him two weeks both times, and each time we got a check for fifty dollars. The more I herded Mrs. Corcoran's cows, the more I didn't like it. As the pasture dried up, the cows made more trouble about trying to get into the alfalfa fields, and as they got skinnier and skinnier Mrs. Corcoran kept blaming me and saying it was because I brought them in too early, or because I didn't graze them where the grass was best. Fred Aultland said it was because I didn't let them get into the neighbors' crops enough to suit her.

Just before school opened she gave me fits because I brought them back to the corral one night at five minutes before six. When she pinned the thirty-five cents into my shirt pocket, she told me that I hadn't earned half of it, and she was only giving it to me because we were so poor. We weren't poor, and I told her so, and yanked the pin out and threw the money right down by her feet. After that she wasn't so mean, and picked it up and passed it to me after I got on Fanny.

I was mad all the way home. When I got there Mother was feeding the hens and turkeys out beside the barn. After I'd pulled the bridle off Fanny so she could go and roll, Mother asked me what the matter was. I remembered what Father had told me about forgetting what Mrs. Corcoran said and not telling anybody, so I told Mother I was mad because I didn't think I was getting paid enough for herding the cows. She put her arm around me and pulled me up against her. Then she patted me on the head, and said, "Son, if you amount to as much as I think you're going to, some day you'll kick on a *dollar* and thirty-five cents a day." I did tell Father about it that night when we were milking, though. And from then on I never herded Mrs. Corcoran's cows.

School started about the first of October. Muriel was old enough to go that year, but she wasn't strong enough to walk the mile and a half, so Father let us drive Fanny. It wasn't a bit the way starting school had been when we first came there. All the kids knew we had a horse now, and that I had ridden up to the mountains to get Two Dog, and that I had made Mrs. Corcoran pay me thirty-five cents a day for herding her cows. They knew, too, about Father fighting to get the irrigating water and about his fixing Fred Aultland's stacker so as to make the hay fall where they wanted it. Everybody called me

Spikes, and Freddie Sprague gave me half an apple at morning recess.

Mr. Lake was the chairman of the school board. They said he always came for the opening day, and he always rode his old white mule. He was a little man—quite a lot older than Father—and he had big joints at the knuckles of his hands. All morning he sat up on the little platform by Miss Wheeler's desk and watched everything we did. While he was watching us, he kept pulling his fingers, one at a time, until he made the knuckle pop. Just when you didn't expect it he would point at somebody and ask him to bound California, or what body of water the Mississippi River emptied into, or something else. He got me on the worst one. He pointed his finger right at me and said, "You! Little tow-headed fella! Go to the board and write me: 'Pare a pear with a pair of scissors.'"

The only two kinds I knew about were *pear* and *pair,* and I got all mixed up on whether there were two *s*'s or two *z*'s in *scissors.* He banged his hand down on the desk and told Miss Wheeler she wasn't a very good teacher, or I'd know better than that. Then he told her to put me back in the first grade in spelling till he came again. I was pretty much ashamed of myself, because we liked Miss Wheeler and I didn't want to get her in a mess with her boss, but Grace got mad. She jumped right out of her seat and told him that it wasn't Miss Wheeler's fault, because we were new there—and that I never could spell *cat* without a *k*, anyway. All the good it did was that he made her stand with her face in a corner till noontime. He said that would "learn her not to sass her elders."

Everybody was talking about old Mr. Lake while we

were eating our lunches, and Willie Aldivote dared me to sneak out in the afternoon and put a burdock burr under his saddle. I pretty near lost my nerve, but the more I looked at him, the madder I got, so halfway between recess time and four o'clock I put up two fingers, and Miss Wheeler nodded at me.

Mr. Lake had a two-cinch saddle, and I only had to loosen the back one a little bit so I could get the burr well up under the middle. From then till school let out, I was so nervous I could hardly think at all, but he didn't make me answer any more questions, so I don't think he noticed me.

As soon as Miss Wheeler tapped the school's-out bell we all grabbed our caps and coats and ran for the carriage shed. The old white mule was tied away over at the east end of the shed, so the boys made a big piece of work about getting their harness down and getting the straps straightened out. All the girls knew about the burr, too, and they stood around twittering and giggling and trying to look as if they didn't see Mr. Lake when he came out and put the bridle on his mule.

Just as soon as he put his foot in the stirrup the old mule went crazy. Mr. Lake let go of the reins and sat ker-plunk down against the board fence, and the mule bucked so hard he'd have made a bronco look like a carriage horse. After he had the saddle slewed way over on the side of his belly, he shot right out through the turnstile gate and raced off up the road. As he went through the gate he smashed the turnstile all to pieces and ripped the saddle off. Maybe Miss Wheeler wasn't a very good teacher, but Mr. Lake was. I learned at least a dozen new words from what he said about that old white mule. I was still shaking from being nervous, but Rudolph Haas was as cool as watercress. He went over and helped Mr. Lake get up and asked him if he couldn't drive him home in his buggy.

Of course, Grace had to tell Mother all about Mr. Lake coming to school. While we were eating supper she told about my not being able to spell "Pare a pear with a pair of scissors," and about her being able to bound California, but she didn't mention having to stand with her face in the corner or my putting the burr under Mr. Lake's saddle. The next I heard of it was three or four nights later when we were out milking. Our Holstein cow's tieup was nearest to the barn door, then came the brindle, so that Father's back was toward her when we were milking. Everything was quiet in the barn, except for the music milk makes when it goes singing down into the buckets, and I was thinking about Two Dog, when Father said, "It's a dangerous thing to put a cockle burr under an old man's saddle. Mr. Lake might have been badly hurt."

I just said, "Yes, sir," and Father never mentioned it again.

I didn't have any more trouble at school for nearly a month—except for my glasses and the cellar door. I don't know why we had a cellar door, because there wasn't any cellar, but we did have one. It was one of those bulkhead doors that slant like a lean-to roof. Some of us were sliding down it one day, when I ran a big, long splinter into my behind. It broke off inside the skin and there was nearly an inch of it in there. Willie Aldivote tried to get it out with the little blade of his jackknife, but he couldn't, so he called Grace. She tried to get hold of it with her fingernails, but she didn't have any more luck than Willie. Then Miss Wheeler picked at it with a needle, and finally she sent Grace and Muriel to take me home.

I think Mother always did kind of like operations. She put a clean sheet over the kitchen table, dropped scissors, darning needles, and Father's whisker tweezers into a basin of boiling water, and rolled up her sleeves. It looked as though she were really going to do a big job, and I wasn't very happy when I shinned up on the table.

It seemed as if Mother were trying to dig clear to China. First she tried tweezers, and then she tried darning needles, but the splinter was so rotten that all she could do was nibble away at the end of it. The more she dug, the more I bled, and the louder I yelled. Grace stood by with a strip of torn sheet to mop off the blood. Every few minutes she'd mop away my tears with the same rag, and tell me that the worst would be over soon. I peeked over my shoulder once or twice, and Mother's mouth was clamped up tight. It was a long operation. I must have been on the table half an hour, but it seemed like a month. At last Mother put both hands on her hips, and said, "Well, we'll just have to let Nature take her course. It will fester in a day or two and come right out by itself."

While she was court-plastering a patch on my behind and helping me get my overalls back on, she explained to me that Mother Nature was the best surgeon of them all, and that everything would work out nicely in a couple of days. All the time she was telling me, I was wishing she had thought of it sooner, and not tried to give old Mother Nature quite so much help when she didn't need it.

That splinter bothered my riding for a week or two while I waited for the fester to come and the splinter to go, but nothing happened, except that a hard little lump formed around the splinter. Once in a while, if I slide around quick, I remember that it's still right there.

I don't know that I should tell about my glasses, because my Heavenly Father and I are the only ones who know how I lost them. Maybe the sun was too bright when I was out herding cows, or maybe I got too much hay dust in my eyes during stacking time, but anyway, as soon as I got back to school my eyes started smarting every time I read very long. Miss Wheeler went to talk with Mother about it, and Mother took me in to Denver to be fitted with glasses. It was lots of fun to peek through the little gadgets the man put over my eyes, and to have drops put in them that made everything seem as if I were looking through a rain-covered window, but I didn't like the glasses. They were silver-bowed glasses, and Mother told me I'd have to wear them all the time, and couldn't do anything where I'd be apt to get them broken because they cost two dollars and a quarter.

I wore them to school three days. That is, I wore them every time somebody caught me and made me put them on. The rest of the time I kept them in my blouse pocket. Miss Wheeler was as bad about them as Mother, and I think Grace was worst of all. I stood it just as long as I could, and then, in the middle of the afternoon of the third day, I put up two fingers. Whoever built the privies at our schoolhouse dug the holes good and deep. When I came back to the schoolroom I had lost my glasses. Miss Wheeler had everybody stay after school to help me hunt for them, but we never found them anywhere. Before we could spare another two dollars and a quarter my eyes got better.

16

A Good Month, with No School

FREDDIE SPRAGUE GOT THE MUMPS IN LATE OCTOBER, and they closed the school. That month was one of the best I ever had in some ways. It started off bad because Bill, our old white horse, died. Father let me go up to Bear Creek Canyon with him to get another load of fence posts. We drove Bill and Nig, and while we were loading the poles it started to rain and sleet. It didn't hurt Nig a bit, but from the time it started, Bill humped his back up like a sick old cow. We had to stop and rest him so often that we didn't get home till long after dark, and that night he died in spite of all the doctoring Father could give him.

Mother wouldn't let us quit studying just because the schoolhouse was closed, so as soon as the supper dishes were washed we had to get our books out. One evening Fred and Bessie Aultland came over to play whist with Father and Mother before we had our arithmetic done, so they sat and talked till we got through using the table. Fred said, "For God's sake, Charlie, don't you know me well enough yet to let me lend you a horse? You could do me a favor by taking that three-year-old I bought at the auction, and gentle-breaking him for me this winter."

"Fred, I couldn't expect a brother to do the things you've already done for me and my family. No, Fred, I can't take your colt. My record for losing horses must be the worst in the country—50 per cent in a year."

Fred slapped his leg and laughed when Father said that. Then he said, "Those nags were 90 per cent dead when you got hold of 'em. A man's just throwing his money away to buy that kind of plugs. They eat just as much as good horses and you can't get any work out of them. I'll bring the colt down in the morning."

Fred brought the big bay colt right after we got done eating breakfast the next morning. He was a beauty, but Father wouldn't let me go near him at first. He tied him up at the far end of the barn and gave him two quarts of oats morning and night, while Nig and Fanny only got peas—vines and all.

The day after we got the new horse Father and I went to Fort Logan with the box wagon. Fanny took Bill's place, but she didn't like it a bit. I guess she had forgotten all she had learned in the spring about working double. She slammed and banged around and threw herself down a couple of times before she decided she was going to have to do it. And all the way down to the Fort she danced and pranced like a two-year-old.

We did our trading with Mr. Green in Logan Town. He had the only general store, but there were nine saloons and a post office, beside the depot. Father had brought a couple of little bags to show Mr. Green. One was beans and the other was peas. There were quite a lot of little beans among them, because they didn't get water enough when they needed it, and some of them were kind of black where the frost had hit them before they were ripe. Mr. Green looked both samples over and said he didn't think he could handle many of the peas, but he'd take all the beans we had—in trade—if they were hand-picked so that we only brought him the full-sized white ones.

Mr. Green and Father talked a long time while I was looking around the store at all the things I hoped we would be able to trade our beans for. Then Mr. Green went into his back room and rolled out three empty barrels. While he was gone to roll out a barrel of flour, I smelled of the empty ones. Two of them had had vinegar in them, and the other one molasses. I ran my finger in through the bunghole of the molasses barrel, and there was still some in there. It tasted good.

You never saw so many groceries as we got that day. Besides the barrel of flour, there were hundred-pound sacks of corn meal, sugar and salt; ten pounds of seeded raisins, and cream of tartar, rice, soda, and saltpeter— and a pound of Baker's chocolate. It seemed we would have enough stuff to keep us fed even if winter lasted clear till June.

That night after supper Father and Mother talked about peas and beans, and did arithmetic problems on the other side of the table from where Grace and I were doing our homework. Father was telling Mother how many square feet of ground he pulled the bean vines from to get the sample for Mr. Green, and the same thing about the peas.

First she borrowed Grace's arithmetic book to find out how many square feet in an acre, and after that she got her marked cup and measured each sample. Then she figured, and figured, and figured. When she was all done, I could tell that both she and Father were the happiest they had been since we came to the ranch.

She had all the answers down on one sheet of paper, and said, "Charlie, we're going to be a lot better off than I ever thought we could be when I saw the leaves on those poor plants curling up in the summer. If I didn't make any mistakes in my figures—and I'm sure I didn't—we'll have a hundred and sixty bushels of beans and a hundred and eighty bushels of peas. Supposing that thirty bushels of the beans are small ones which will only bring four cents a pound, and that thirty bushels are frosted and will only be good for pig food; that would leave a hundred bushels of good ones. At five cents a pound, our share will be worth a hundred and eighty-six dollars . . . that is, if there are sixty pounds in a bushel.

"You know, practically all the peas of the variety we have are used for soup, so it doesn't make a particle of difference if some of them are small—they should all bring the same price. Let's say that will be four cents a pound; our half should bring in a hundred and ninety-two dollars. I can't see any reason why we shouldn't be able to afford a good horse like Fred Aultland's."

Right after breakfast the next morning, Father hooked Fanny to the buckboard, and Mother took all the other youngsters to Englewood to buy stockings and underwear and things. I had a day's work helping Bessie Aultland pick apples, so I left before they did. We picked bushels and bushels of apples, and when Bessie took me home, just before supper, she helped me put two bushel basketfuls on the buckboard for us.

As we came near our house I could see what looked like three big white sacks of grain hanging from a crossbar at the back of our barn. I jumped off while the bays were making their circle in our yard, and ran around the barn. Our three biggest pigs were hanging there dead, with all the hair scraped off them. It kind of startled me at first, and I guess Father noticed it. He came right over and bent down on one knee beside me. Then he put his arm around my shoulder, and said, "There isn't a thing to be afraid of, or to feel bad about, Son. The only time to feel sorry for anything—or any-body—that dies is when they haven't completed their mission here on earth. These pigs' mission was to get big and fat so as to make food for us. They have done a good job of it and their mission is completed. And I do want you to know this: they didn't know what was happening, and they weren't hurt a bit—they didn't even squeal." Father could always explain things like that so I'd under-stand.

Everybody worked on the pork the next day. Father cut the hams and bacon and side meat, while Philip and I stripped all the fat off the insides, and ground up the scraps for sausage. Father made a separate pile of the leanest scraps, and we ground them for mincemeat.

Mother and the girls were just as busy in the kitchen as we were outside. They rolled all the sausage into little cakes the size of a turkey egg, fried them slow, and packed them away in stone crocks; tried out all the lard,

146

and made the livers and hearts into sausage. Then they chopped apples and made the mincemeat. It was stewing on the back of the stove when we came in to supper. I never heard of making mincemeat with pork before, but it smelled and tasted better than any other I ever ate.

We got a lot of things done that month when school was closed. We were the only people anywhere around who didn't have a corral and a dug-out cellar. Mother had been worrying ever since the big wind because we didn't have a storm cellar, and Father had been saying he'd build one as soon as he had time to get to the mountains for poles. I couldn't figure why we needed poles to build a cellar, but I didn't like to ask Father. On things like that, he always used to tell me I could learn more if I kept my eyes open and my mouth closed.

Mother must have mentioned something about wishing we had a cellar half a dozen times while we were packing the barrels of pork away in my room in the bunkhouse. At breakfast the next morning, Father winked at me, and said, "Do you think we could spare time to go up the canyon for a load of poles today?"

Of course, I did think so. And right after breakfast he started putting Bill's harness on the new colt. Then he sent me up to Aultland's on Fanny, and said to tell Fred we'd like to borrow one to fit her. I was so excited about going to the mountains with Father that I didn't think much about what we were going to do with three harnesses, but Fred did. As soon as I told him what I wanted and where we were going, he scratched his head and said, "Has your old man gone loco? If he thinks he's going to harness that green colt and take him up to the

147

mountains, along with Wright's old mare, he's either the bravest man I ever seen, or a damn fool." I told him Father was the bravest man he ever saw, and wasn't any fool, so he let me have the harness.

I had to walk Fanny all the way home, because the harness slapped around so much when I made her canter. All the way, I kept thinking about what Fred had said. I was kind of scared, too, about what would happen when Father got all three horses hooked up to the wagon.

When I got there, he already had old Bill's harness on the colt, but he had it fastened on with three or four extra straps, and the traces were tied up around the back of the breeching. The colt was sweaty and nervous, but he wasn't raising Ned at all.

After Father had hooked Nig and Fanny to the box wagon, and Fanny had got over slatting around, he led the colt out and tied him up close to the back of it. He hitched his head to both sides so that he had to keep it right in the middle of the tail gate. Then I ran to the house for our dinner pail, and we started off. You never saw a horse buck and kick much worse than that colt did when he felt the harness flopping around him, but Father had it strapped on so tight and his head tied up so short that he couldn't hurt anything.

By the time we went past Aultland's house he was soaking wet, but he wasn't bucking any more—just dragging back on the halter ropes and trying to spit out the bit. Fred was standing out in the yard when we went past. When I waved to him, he waved back, and yelled, "I'll take back what I said, Spikes." I just grinned, because I knew all the time that whatever Father did would be right. Father must have guessed what Fred had said, because I didn't tell him, but he looked over at me and grinned, too.

When we had loaded our poles and got down out of the canyon, Father tied the colt alongside of Nig. That time he fastened a strap from the colt's outside trace over onto Nig's breeching so that he couldn't swing his hind end around sideways. At first he'd hang back till the single-tree bumped against his legs, then he'd jump around and kick, but Nig didn't care, and then he learned to stay up where he belonged.

Father unhitched Fanny after we got home, and while we still had the load of poles on the wagon he hooked the colt in her place. By that time he was used to the harness, and I guess he was a little tired, but he hardly made a bobble. In half an hour he was pulling like an old horse.

We hauled poles for three days, and took the colt with us every day. After the first one, Father put him in Fanny's place just as soon as we got down out of the canyon, and from then on he behaved better than Fanny did.

Before we started hauling poles Father had dug a little ditch around a patch of ground in the back yard. He made a trough that ran out there from the well, and every morning and night it was my job to pump the ditch full of water. In three days the ground had softened up in good shape, so we borrowed Carl Henry's slip-scraper and started digging our cellar. I had learned how to ease a horse up into the collar for a hard pull while we were stacking hay. Father hitched Nig and the new colt to the scraper and let me drive them while he held the handles.

If I didn't start the team real easy when Father raised the handles of the scraper, the cutting edge might catch and throw him up under the horses' heels.

Father explained it to me before we started, and I was so afraid I might do something wrong and get him badly hurt that my hands were shaking when I reached out to take the lines. He wouldn't let me take hold of them then. He said I'd have to stop a little while and get my mind straightened out, because a horse could tell through the feel of the reins if the person driving him was afraid. Then he told me I had already proved I could make a horse do what I wanted it to, so there was no reason to be afraid now. It made me proud to hear him say that, and when I reached out for the lines again my hands were steady. I wrapped the reins around them and called, "Get up," with my voice as deep in my throat as I could make it go.

We scooped out a hole nearly as big as our kitchen. While Father dug the corners out square with a pick and shovel, I peeled bark off the poles with his drawknife. It took five days to build the cellar. After the hole was dug we cribbed the walls up with poles like a log house. We made the end walls half round at the tops and then laid poles across to make the roof. Grace and I stuffed all the cracks on the outsides of the walls and roof with straw while Father made the door and the steps. Then we hitched up the horses and, with the scraper at the end of a long rope, filled dirt in tight around the sides and over the roof till it looked like a little hill with a trap door in it.

The next week I peeled poles while Father built them

into a corral. It was a good one, with a six-pole fence five feet high. Father set a big, high post for the gate to swing on. Then he made the gate out of slim poles with the butt ends toward the hinges, and a guy wire running from the top of the post to the lighter end of the gate so it could never sag.

While we were building it I got thinking how lonesome our little house had looked to me, sitting out there on the prairie, when I had first seen it from the hill by Fort Logan. When the last nail was driven and the hasp was put on the gate, I got Father to let me put Nig and the new colt and our two cows in the corral. Then he let me take Fanny and ride up to that hill again, so I could look at our place and see how much it looked like a real ranch now.

17

I MEET THE SHERIFF

ALL DURING THE TIME WE WERE BUILDING THE CELLAR and the corral, Grace and I had to do our schoolwork after supper. Father worked with us, too, but I couldn't make out what he was doing. He had some big sheets of wrapping paper that came with the groceries, and his steel square and dividers, and while we were studying, he'd be drawing pictures. Once in a while he'd ask Mother to figure out an arithmetic problem for him, and then he'd change his drawings all around.

The morning after we finished building the pole corral he and I drove Fanny to Englewood. It was at the end of the Denver streetcar line and had lots more stores than Fort Logan. First we went to the blacksmith shop and got a couple of lengths of angle iron, small pulley wheels, and pieces of round iron rod. Then, at the hardware store, we bought sheets of galvanized iron, three or four kinds of screen wire—some coarse and some fine—and lots of bolts, screws, and other things. There was so much that I knew it would cost a lot of money, and I asked Father if we'd have any left. He took out his long leather pouch and showed me that there was quite a little silver and some bills in it. Then he said that part of it was mine, and asked me if there was something I wanted to buy. I told him I wished I had a steel trap, so we went over to the corner where the guns and traps were, and he helped me pick out one the right size for prairie dogs and skunks.

I was wondering what we were going to do with all the hardware and iron, and after we started for home Father told me we were going to build a winnower. He said it would cost too much to have a big machine come to thresh our peas and beans, but we'd have plenty of time during the winter to do it with hand flails and a win-nower.

After we got home, he spread a roll of brown paper out on the bunkhouse floor, got his drawings, and began cutting patterns the way Mother did for making clothes. Father didn't need me to help him, so I went out to set my new trap. Before I left he told me I'd have to set it quite a ways from the buildings, so King or one of the cats wouldn't get into it, and then I'd have to stay away from it if I expected to catch anything. I took it clear over beyond the railroad tracks and set it near a prairie dog village. I knew they liked peas, so I sprinkled dried grass over it till it was almost hidden, then put a little handful of peas right above the trigger plate.

After it was all set I went back to the bunkhouse and watched Father cut patterns for a while, but I kept asking him if he didn't think it was about time I brought in the cows. I was wondering if I'd caught a prairie dog yet, and I could go around that way when I went for the cows without having it seem too obvious that I was anxious about my trap.

Father didn't let me go for them till sunset. As soon as I got out behind the barn where nobody could see me I ran to beat the band. From the railroad track I could see that there was something in my trap, but it didn't look like a prairie dog. It looked bigger and brighter. When I got close enough I found that it was a big cock pheasant. His head was inside the jaws of the trap and there were a few feathers blowing around from his flapping when it broke his neck.

The first thing that popped into my head was what Fred Aultland had said about spending the rest of your life in the hoosegow if you killed a pheasant. I was so scared I got all shaky. First I thought the best thing to do would be to get him out of there and hide him in the bottom of a deep gulch. I looked all around to see if anybody was in sight, then I stepped on the trap spring and took him out. The jaws had pulled a lot of feathers out of his neck and had almost bitten his head clear off. So, if I hid him in a gulch and somebody found him, they'd know just what had happened. Then I figured that if I didn't hide him, but just threw him down in the gulch, the coyotes would come and eat him.

It wasn't very dark yet, and I was afraid somebody might see me if I just lugged him away across the prairie, so I took off my coat and wrapped him up in it. After gathering up all the loose feathers I started for the gulch, but the farther I went the more I worried for fear the coyotes might not eat him. It seemed as if it would be like trying to eat a pillow. I was sure they wouldn't do it because they'd get their mouths all full of feathers. Then I thought if I picked him they'd be sure to eat him—and I could let the wind blow the feathers away so nobody could ever tell I'd had anything to do with it.

By that time I had reached the edge of the gulch and slid down over the bank to start picking. It was getting pretty dark, but when I unwrapped him I could see what a mess I was in. The pheasant hadn't bled a bit with the trap jaws around his neck, but after I had wrapped him up, blood had run out of his mouth till the inside of my coat was all red and sticky. I didn't know what to do. Just as if he were deciding it for me, a coyote howled from somewhere farther down the gulch. I bundled the pheasant up quick and went after the cows. I knew I'd have to have Father's help to ever get out of the mess I was in.

Everybody was in to supper by the time I got the cows home, so I hid the pheasant, slammed the bunkhouse door as if I'd been in to hang up my coat, and went into the house. The minute I stuck my head inside the kitchen door, Mother said, "What in the world have you been up to now? You look as though a ghost had been chasing you."

I said I hadn't been up to anything, but the cows had been way over by the big gulch and it was all full of coyotes, and maybe that had scared me just a little. Then she told Father that I was too young to be way off out there after dark, and that I'd have to start after the cows earlier. He just said, "mmhmm," and bowed his head to say grace as soon as I was in my chair. He might just as well have said to me, "You and I will talk more about this later."

We went out to milk right after supper. I don't think I had more than a dozen squirts of milk in the bottom of my bucket—just enough so that it didn't ring any more—when Father said, "What did you do, get your own foot in your trap?"

I said, "No, sir." Then I went ahead and told him about catching the pheasant, but I didn't tell him about wanting to hide it. I asked him if he thought they'd put me in the hoosegow, as Fred said, if the sheriff found out about it.

Father didn't say a word for a minute or two. Then he said, "It isn't a case of 'if the sheriff finds out about it.' It's a case of your breaking the law without intending to. If you tried to cover it up, you'd be running away from the law. Our prisons are full of men whose first real crime was running away because they didn't have courage enough to face punishment for a small offense. Tomorrow you must go to see the sheriff. I'll explain to Mother about your coat."

I didn't have a very good night. I couldn't keep my mind on my business after supper, and Mother nearly spanked me because I got all mixed up and couldn't say the table of twelves. She gave me a glass of warm milk before I went to bed, but it didn't make me sleep any better. Whenever I wasn't awake I was dreaming. Mother used to recite "The Ballad of Reading Gaol," and that night I kept dreaming I was the man in the ballad. Every time I'd wake up in the pitch dark, I'd put my hand over and feel for Philip to make sure it wasn't really so.

After breakfast I begged Father to go to Fort Logan with me to see the sheriff, but he said, "No. You haven't learned to ask for advice before you get into scrapes, and it isn't fair to expect help in getting yourself out every time."

I told him he'd have to go with me, because I didn't even know who the sheriff was and I'd never be able to find him alone.

He just boosted me up on Fanny's back and handed me the bag with the pheasant in it. Then he said, "You found Two Dog's lodge all right without any help, didn't you? If you ask at the post office, I think they'll be able to tell you where you'll find the sheriff."

I'd always cantered Fanny all the way to Fort Logan—and right up to the hitching rail—but that morning I made her walk. For a while I thought about running away, but the only place I knew to go was up to Two Dog's. If he'd lived alone, I guess I might have gone, but I remembered what a hurry Mr. Thompson had been in to bring me right home, so I decided I'd better not try it. All the time I was thinking about running away, I kept getting a squirmy feeling in my stomach because of what Father had told me when we were milking. After I made up my mind that I was going to be brave and face the music it stopped a little, but it came right back again when I went into the post office.

The lady behind the window told me I'd find the sheriff over at the Last Chance Saloon, just outside the gates of the Fort. At first I thought that would give me a good enough reason to go home without seeing him, because I knew what Mother thought of saloons, and of course she wouldn't want me to go into one. So I climbed back on Fanny and started down the street toward the Morrison wagon road.

I knew Mother would say I had done just the right thing, but I tried not even to think about what Father might say. I couldn't help it, though. And I wasn't a bit sure he wouldn't say it was running away from the law and tearing boards off my character house. We had just turned into the Morrison wagon road when I got a big lump in my throat. Then I pulled Fanny around and galloped her back to the hitching rail in front of the Last Chance Saloon.

My heart was thumping like sixty when I went in through the little swinging doors. I was scared, but I was a little bit proud, too, that I had business big enough so that I could go right into a saloon.

I stopped just inside the doors—it was kind of dark in there and there were about a dozen soldiers and other men leaning against the bar and talking loud. The man behind it yelled, "What you doing in here, Bub? Who you want to see?"

He leaned across the bar, and I went over and told him I wanted to see the sheriff. He just jerked his thumb toward the back of the room and said, "The big fella."

The sheriff was talking to another man when I got back there, so I stood behind him and waited for him to get finished. He was the biggest man I'd ever seen; my head didn't come up as high as the top of his cartridge belt, and the longer I waited the bigger the lump in my throat grew. At last the man behind the bar came back and told the sheriff I was there, so he leaned over and said, "What can I do for you, Son?"

I had to swallow hard before I could make a sound, then I said, "I broke the law and Father made me come down to tell you."

He said, "Well, well, well! We'll have to look into this." While he was saying it he sat me up on the bar in front of him and asked me what I'd done.

All the men along the bar came and made a big crowd around us, I showed him the pheasant and told him that I didn't kill it on purpose, but it got in my trap when I was trying to catch a prairie dog.

He took the pheasant and laid it on the bar beside me. Then he rumpled up all its feathers and felt it all over with his hands. After he'd finished, he said to the men, "By God, that's the way he got it all right. I'd 'a' sworn his old man shot it and sent the kid in to get himself out of a pickle."

I didn't like that, and I guess I must have yelled, "Father would not try to get himself out of a pickle."

Everybody laughed and hollered, and the sheriff said, "Kind of like your old man, don't you? What makes you think he wouldn't try to get out of a scrape?"

I told him, then, what Father had said about our prisons being full of men who ran away from the law, but that time nobody laughed. The sheriff put the pheasant back in the bag and handed it to me. He said the law was that you couldn't shoot a pheasant, but he didn't remember anything in it against catching one in a steel trap, so I'd better take it home for Mother to roast.

Then he asked me if I'd like a drink. I told him I liked brandy with sugar and water in it, but Mother would only give it to me when I got blue from the cold. All the men laughed some more, and one soldier yelled, "Set the kid up a shot of brandy, Tom." But the sheriff shook his head, and told the bar man to make it birch beer. At first I didn't know if Mother would want me to drink it, but the sheriff said it was all right. It was, too.

I cantered Fanny home as fast as she could go, and Father didn't scold me for bringing her in hot. Everybody came out to the barn when I got there, and I told them what the sheriff said, but I didn't tell them about the saloon and the birch beer. Grace and Philip seemed to think I was a lot more important now that I had talked to a real sheriff. Mother took the bag with the pheasant in it, and said the sheriff must be a fine man. I guess Father thought so, too, because he said he must look him up the next time he went to Fort Logan. I pretty nearly told Father where to find him but I caught myself just in time.

That night when we were milking, he told me it had been a day I should remember. He said it would be good for me, as I grew older, to know that a man always made his troubles less by going to meet them instead of waiting for them to catch up with him, or trying to run away from them.

159

• • •

The next morning Fred Aultland came for Father right after we finished breakfast. I guess Mother knew he was coming, because she had already told Grace and me we could spend the day visiting Willie and Etta Aldivote. I tried to get Father to tell me where he was going, but all he would say was, "Oh, we're going way over by Littleton to see a fellow about a dog." When Father said that it always meant he wasn't going to tell you where he was going, so I didn't ask any more.

Grace and I had a fine time at Aldivote's. Their house was a soddy on the front, and was dug right back into a bank like a cave. They had a big barn full of hay, and a donkey and half a dozen horses we could ride. The girls made us play house with them some of the time, but we made them try to ride the donkey, and they took some of the craziest spills I ever saw. I think we had the most fun, though, jumping from the high haymow down onto a pile of straw on the barn floor. It must have been thirty feet.

Father and Fred got home just a little while after we did. They were leading a beauty of a bay horse. He wasn't quite as big as Fred's new one, but he had a lot more ginger. I guess we youngsters and Mother frightened him when we all ran out to look. But, anyway, he started bucking and pitching like Old Harry just as they were coming into the dooryard. If he hadn't had a half-inch rope around his neck, as well as having a halter on, I think he would have broken away. They had quite a time putting him into the corral without getting kicked. And as soon as they let him loose, he went racing around and around the fence, kicking his heels higher than the top rail.

Mother was all flustered about how dangerous the new horse was, and made us children stay in the house till suppertime. While we were eating, Father began telling what fine horses there were at the auction, but Mother asked, "Didn't they have any nice gentle horses, Charlie? I'm afraid one of these bad ones is going to hurt you."

Father grinned and said, "He isn't a bad one, Mame. He's a good one. One of the best I ever saw. But he's just a three-year-old, right in off the range, and he's never had a man's hand on him until today."

It looked for a minute as if Mother were going to cry. "I don't care if he's three or thirty. He's bad; bad all the way through, or he wouldn't have thrashed around and fought the way he did. I don't see why you didn't buy a nice gentle horse. You can't tell what will happen to Ralph with that kind of a horse around the place."

Father got up and went around behind Mother's chair. He put both hands on her cheeks and patted them. Then he said, "Son, I want you to make me a promise that you won't go near the new horse, or into the corral while he's in there, until I say you may."

My mouth was full, but as soon as I could swallow, I said, "I promise, Father."

"I could have bought a gentle horse, Mame. There were some, not much better than Nig, that brought seventy-five dollars. That's what this one cost, and that's all we planned we could spend. Well-broken, young horses, as good as he is, were bringing a hundred and a quarter or more. I picked him out before they ever put a rope on him, and there's nothing vicious about him; he's simply crazed with fear right now. I'll be very careful in breaking him."

Mother reached up and patted his hand against her face. "You will be awfully careful, won't you, Charlie?" was all she said. Father was real good about making people believe what he said.

18

FATHER AND I BECOME PARTNERS

WHEN WE LIVED IN EAST ROCHESTER, MOTHER USED to let Grace and me take the money to pay the grocery bill every Saturday. Mr. Blaisdell always gave us a little bag of candy when we came in to pay, but since we had moved out to the ranch we never got any. I liked all kinds of chocolate, but I liked the bitter kind Mother baked cakes with best. The last Christmas before we came west, she had made fudge with some of it. It was the best candy I ever tasted. I got thinking about fudge, and one night I asked her when she was going to make some more. She said maybe she'd make some when Christmas came, but sugar cost too much to be using it up in candy we didn't need.

The more I thought about fudge, the more I thought about the bar of Baker's chocolate we got with our last groceries, and the more I wanted some of it. Baked beans, pea soup, and fried sidemeat had tasted all right before, but thinking about chocolate, they didn't even make me feel hungry.

The next afternoon when I was helping Father on the winnower, I was thinking of what he had said about going to meet your troubles and how much less they would be. I don't know if I'd even stopped thinking about that when I began daydreaming about chocolate again. It was right then I got the idea: If I should whack a chunk off the end of that bar of chocolate, Mother would be sure to miss it. Then, before she had any idea who had done it, I could confess and probably wouldn't even get a spanking for it, any more than I did for going up to Two Dog's.

I waited till she was out feeding the chickens, then told Father I was thirsty and thought I'd go in for a drink of water. All the time I was going into the house and getting the bar of chocolate down out of the cupboard, my head kept wanting to think about tearing boards off my house, but I wouldn't let it, because I told myself that was only when you did things you shouldn't and then lied about it. I wasn't going to lie at all about the chocolate.

I heard Mother coming just when I had the knife ready to whack off the end of the bar, so I had to slip it into the front of my blouse and pick up the water dipper quick. Before I went back to help Father I went to the barn and hid the bar of chocolate back of the currycomb box.

All the rest of the afternoon, I didn't like to look at Father. I tried to get him to let me go over to see Willie Aldivote, but he wouldn't. Every time he spoke it made me jump, and my hands got shaking so I couldn't hold the pieces still enough for him to solder. He asked me what was the matter, and I told him it was nothing except that my hands were getting cold. I knew he didn't believe me, and every time he looked my way my heart started pounding, because he could always tell what was going on inside my head. It seemed it would never come time to go for the cows. I didn't want the chocolate any more; I just wanted a chance to put it back without being caught.

On the way out for the cows, my heart stopped pounding so hard, and I could think better. I hadn't really stolen the whole bar of chocolate, because I had only meant to take a little piece, and that's as much as I would have taken if Mother hadn't come in just when she did. If I put back the whole bar, I wouldn't have done any-thing wrong at all. I'd nearly decided I would do it, but just thinking so much about chocolate made my tongue almost taste the smooth bitterness of it. It didn't seem as if it would be very wrong if I only took a small piece. Then I got thinking that if I took a sharp knife and cut about half an inch off the end—with a good clean slice—Mother might never notice it.

I was nearly out to where the cows were picketed when I remembered what Father had said when I got my trap: some of the money in his pouch was mine because I had earned it. Why wouldn't it be all right to figure that the bar of chocolate had been bought with my own money, and in that way I wouldn't be stealing it at all. That seemed to fix everything, and I got planning how I would go out to the barn every night after school and whittle off a little piece of chocolate.

I could have felt all right about the whole business if it hadn't been for Mother's reading. Sometimes, on

Sunday afternoons, she used to read just to Father, but any of us could stay in the house and listen if we wanted to. He often had her read Shakespeare's plays, and the one he liked best was about Hamlet. I liked it, too, and used to listen every time she read it.

I had just pulled the picket pins and was heading the cows home when the bad king's prayer came into my head, and I couldn't get it out. I tried to think about how Hi dived off his horse and came up on his feet, and about Two Dog, and King, and everything else, but my head kept on saying, "Oh, my offense is rank," until I thought I'd go crazy.

We were nearly to the railroad track when I decided to leave the whole matter to the Lord, and twisted out a dried soapweed stalk with seed pods on it. When you slung one of them up in the air it would wobble and twist all around so that you never knew which way it would come down. I told myself that if it came down with the pods to the west I'd take the whole bar of chocolate back. If it came down pointed to the south, I'd take half an inch off the end, but if it came down pointed to the east, it had been bought with my own money and it wouldn't be stealing to keep it.

I swung the pod stalk around my head a few times and flung it as high as I could, then I shut my eyes tight till I heard it land. When I opened them the pod end of the stock was pointed almost toward the west, but not quite. It was a little bit toward the south.

There was a bright moon when I went to bed that night, and it was sharp and frosty. I couldn't go to sleep and kept trying to remember how much the pod end of that stalk had really been pointing toward the south. At last I heard Father put King outside for the night, and a little later when I peeked under my curtain I could see that he had blown out the lamp.

I pulled my overalls up over my nightgown and took my shoes in my hand. After I was out in the yard I slipped them on and took the axe from the chopping block. It was good and sharp, and I was sure I could peel off a smooth, thin slice of chocolate with it.

It was dark as tar inside the barn, but I felt along the wall for the currycomb box, and lifted the chocolate box out from behind it. King had followed me, and I nearly fell over him when I was groping for the door, but it was so light outside that you could almost have read a book. I shook the bar out of the box, unwrapped it, and laid it on the lower rail of the corral fence. Just as I was starting to cut it with the axe, Father said, "Son!"

I couldn't think of a thing to say, but I grabbed up the bar of chocolate and shoved it inside the bib of my overalls before I turned around. He picked me up by the shoulder straps—just as he'd have picked up a kitten that had wet on the floor—and took me over to the wood pile. I didn't know anybody could spank as hard as he spanked me with that little piece of board. It felt as if my bottom were going to catch fire at every lick.

Then he stood me down and asked me if I thought I'd deserved it. He said it wasn't so much that I took the chocolate, as it was the way I took it, and because I tried to hide it when he spoke to me. But it was the next thing he said that hurt me worse than the spanking.

He said, "Son, I realize a lot better than you think I do

that you have been helping to earn the living for the family. We might say the chocolate was yours in the first place. If you had asked Mother or me for it, you could have had it without a question, but I won't have you being sneaky about things. Now if you'd rather keep your own money separate from the family's, so you can buy the things you want, I think it might be a good idea."

I never knew till then how much I wanted my money to go in with Father's. Ever since we bought the cows, I had been able to feel I had a part in all the new things we were buying to make ourselves real ranchers, and it looked as though it were all slipping away from me. I had felt I was beginning to be a man, but I guess I was still just a baby, because I hid my face against Father's stomach and begged him to let me put my money in with his.

Father hadn't been coughing nearly so much that fall as he used to, but he coughed and it seemed as if he choked a little before he answered me. He said he didn't want a sneaky partner, but if I could be open and above-board he didn't know a man he'd rather be in business with.

I couldn't help crying some more when he told me that; not because my bottom was still burning, but just because I loved him. I told him I'd never be sneaky again, and I'd always ask him before I did things. We walked to the house together. At the bunkhouse door he shook hands with me, and said, "Good night, partner." When I went to sleep, my hand was still hurting— good—from where he squeezed it when we shook hands.

19

Trapping Pheasants

Wʜɪʟᴇ ᴡᴇ ᴡᴇʀᴇ ɪɴ sᴄʜᴏᴏʟ Fᴀᴛʜᴇʀ ʜᴀᴜʟᴇᴅ ᴀʟʟ ᴛʜᴇ beans in from the field and made a stack right beside the barn. Every day he would flail out a big pile of them, and when I got home we would winnow them out. The vines were so musty that the dust nearly choked us and it made Father cough terribly. And the beans weren't very good either. Almost half of them were little tiny ones, and a third of all we winnowed were black from being frozen before they were ripe.

Mother would come out every night when we were finished. Then she and Father would look at the bags of beans in the barn and at what was left of the stack. He'd say, "There are still quite a few left in the stack, and they're from this end of the field. That's where I took the samples we figured from, and I'm in hopes they'll run a little better."

Mother would bite her underlip in between her teeth, and then she'd say, "Don't worry about it, Charlie. We'll get along all right—one way or another. I think it's the worry as much as the dust that's running you down so. Why don't you have Mr. Lewis come with his machine and finish the threshing for you? We could pay him with part of the peas and beans, couldn't we?"

Then Father would put his arm around her, and they'd walk to the house while he told her that, with a wife like her, a man had nothing to worry about; and she'd tell him that the Lord had always provided for us and that He always would.

Father left our new horse out in the corral all the time we were threshing beans. But every evening he'd take him in a few oats in an old bucket. At first the colt wouldn't come near him, but crowded into the farthest corner of the corral. Then Father would set the bucket down and come outside the gate. After a while, the colt would start sticking his nose out toward it. Pretty soon he'd creep up and grab a mouthful, jerk his head up quick, and watch us while he chewed them. Every day he seemed to be less afraid, and the last day of bean threshing he came trotting right up to the gate when he saw Father coming toward him with the bucket.

The peas were easier to thresh than the beans, and weren't quite so musty, but there were an awful lot of small ones. Father made me a little flail out of an old broom handle and a singletree stick, and let me stay home from school to help him thresh. The only way Mother would let us do it was with wet cloths tied around our faces. Maybe it was a good idea, because we didn't breathe in so much dust, and the wet cloths got so cold that we had to flail like sixty to keep from freezing.

As soon as we opened the stack and started threshing peas, the pheasants would come every morning at daylight. There were as many as a dozen on top of the stack one morning when we went out to milk. Father said they were getting to be pests, and would rob us of ten pounds of peas every morning.

While we were milking I got thinking about all the peas the pheasants were robbing us of, and about how good the one Mother roasted had been. That night I set my steel trap right in the middle of the open place on top of the pea stack. The next morning there was a nice fat cock pheasant in it. At breakfast Father and Mother talked about whether or not it was all right for me to have done it. At first they said it was against the spirit of the law for me to catch him, but I told them again what the sheriff said about there being nothing he knew of in the law against catching pheasants in a steel trap.

Father said, "You know, Son, a man sometimes has to consider the spirit of the law as well as the actual words."

But all Mother said was, "Wasn't that other one delicious?"

I kept wondering all day about trying to trap another pheasant. Father hadn't really said I couldn't, but he hadn't said I could, either. I started to ask him two or three times, but without Mother there to say how delicious the first one was I thought I'd better not. Then I thought I'd just slip out when I went to bed and set the trap. If I didn't catch anything, of course, Father would never know anything about it, because I was the one who always climbed up on the stack to pitch the vines down. If I did catch one, Mother'd probably say, "How delicious!" again, and that might be all there'd be to it.

I thought I had my mind all made up, but I tried to keep my back turned toward Father as much as I could, so he wouldn't be able to see what I was thinking. Then I'd get worried that he might be able to see, anyway, and I'd start remembering about our being partners, and the chopping block, and how good my hand felt that night after he shook it. I tried to tell myself it wouldn't be sneaky to set my trap without his knowing about it, because he didn't always know when I set it for prairie dogs, but my head kept saying, "It would, too, be sneaky!"

I didn't figure out what to do till we were eating supper; then I said to Father, "Do you think I ought to drive a stake down in the pea stack to keep the pheasant from flying off with my trap?"

Instead of looking at me, he looked up at Mother. We were having spareribs and beans for supper that night. She was helping Hal get the meat off the bones when Father looked up at her. I don't know whether she saw him or not, but she kept right on cutting Hal's meat, and said, "I do hope the children won't get tired of pork and beans before spring comes."

Father looked down at his plate again, and said, "It might be a good idea, Son."

I got a pheasant off the pea stack every morning till we finished threshing. I guess that made up a little bit for how few bags of peas we got out of it.

The day we finished winnowing, we carried all the peas and beans into my room in the bunkhouse. Then we measured them all out into other sacks. When we were through putting two bushels into each one, we had just an even hundred sacks. Forty nine of them were peas, twenty-eight were little beans, and twenty-three were large ones. Of course, the frozen beans were still in with the good ones, and there were lots of them. Father kept running the big ones through his hands, and saying, "They'll be more than half salable."

• • •

I had to go right back to school again just as soon as we were done threshing, so I didn't get much chance to see Father break the new horse. He knew how disappointed I was, and told me I could be the one to name him. It seemed only right, since he was taking old Bill's place, that he should have his name, too, but he was so young I didn't want to call him "Bill," so I named him "Billy."

Billy had gentled down enough before I went back to school so Father could lead him with a halter, and he was being tied up in the barn every night. By the end of the week, he was hitching him up with Old Nig, and pulling a load of dirt around in the field. The first time I saw him hooked up, he was still trying to run away from the wagon, but Father had on a heavy load, and the brake set, so all Billy could do was jump and pull.

About the only things we did the rest of the winter were to go to school and pick over beans. Right after supper every night, Father would pour a big pile of them in the middle of the kitchen table. Then Mother would read to us while we all sat around the table with pans in our laps and sorted the good beans from the frozen ones.

One Saturday Father and I threshed and winnowed a big sack of the oats that we had raised with our alfalfa. It took a lot of threshing, and he had to turn the crank on the winnower fast, because there was so much hay and straw and so few oats. The next Monday he took them to the mill at Littleton and came home with the sack half filled with oatmeal.

20

THANKSGIVING AND CHRISTMAS

THE MORNING BEFORE THANKSGIVING FATHER PUT three sacks of peas on the back of the buckboard, and he and Mother started for Denver before Muriel and I went to school. We had to walk because they drove Fanny, but Grace stayed home to take care of Philip and Hal.

It was after dark when Father and Mother came home, but I heard Fanny's feet when she came over the bridge at the gulch. She was just walking, and King and I ran down the wagon road to meet them. I thought it would be fun to jump out and frighten them, so I flopped down behind a big tumbleweed and held King close up against me. Before they were opposite us I could hear Mother talking. "Let's not allow the small price we got for the peas to spoil our Thanksgiving, Charlie. With five healthy children we have more to be thankful for than most anyone I know. And we have enough to feed them till spring, even if there won't be much variety."

Hearing her talk like that when she didn't know I was there made me feel like I was being sneaky, so I jumped up and yelled, "Hi, Father!"

We did have a good Thanksgiving, too. Father and Mother must have sat up till nearly midnight to get things ready. They didn't let us look into the box of groceries when they got home, and made us go to bed early. But when we got up in the morning, our biggest turkey was all dressed and hanging up near the kitchen door to chill. At breakfast Mother said, "Grace and I have a lot of work to do this forenoon. I want the rest of you to get bundled up and stay right out from underfoot till dinner is ready."

That was the first time Father let me drive Billy. The section hands had been putting some new ties in the railroad track, and had left the old ones so we could have them for firewood. Father wouldn't let me hitch Billy to the wagon, but said I could lead him out of the barn. Then, after I had hooked Nig's traces, he passed me the reins.

Billy still tried to run away sometimes, and I had to be real careful that my hands didn't shake a bit, so he would know I was a little mite afraid. I didn't try to sit on the seat, but stood down on the wagon bed where I could brace my feet in good shape. I guess Billy knew all right that somebody besides Father had hold of the lines, because he started off dancing and hopping. But I pulled hard on the reins so as not to give my hands any chance to shiver. And by the time we got out where the ties were he was behaving pretty well. Every time Father heaved a tie onto the wagon Billy would jump, but he didn't try to run away, and he pulled just as well as Nig when we were going back to the house.

I was nearly starved before Mother came to the door and called, "Dinnnn . . . nnnerrr!" And you never saw such a dinner in your whole life. There were sweet potatoes and white potatoes and boiled onions, and squash and turnips and cranberry jelly, besides the turkey. When that was gone, there was mince pie and pumpkin pie; and afterwards a pound of cracked nuts . . . and a plate of fudge. We all ate so much we could hardly get up from the table. Then Father and all of us lay on the floor by the stove while Mother read us "Snowbound." I think it was about the best day any of us had ever had.

The only other thing that happened before Christmas is one I don't even like to remember about. Since we moved to the ranch, Father had spent all his spare time setting fence posts. Soon after Thanksgiving he set the last ones, so that he had a row clear around the whole place. The Saturday before Christmas, we started stringing the secondhand barbed wire we had bought from Mr. Cash.

Father had bought a wire stretcher that worked kind of like a pump. The more you pumped the handle the tighter the wire got. Philip came out to watch us, but Father wouldn't let me do anything except bring him staples. And he told me to keep Philip way back away from the wire till he had it stapled tight, because it might break and hurt us. Father had just finished stretching the top strand of wire when I noticed a big bald eagle. He seemed to be about a mile high and was almost standing still up there. I forgot all about Philip and the fence and everything else, and was thinking of all the things I could see if I were sitting up there on the eagle's back.

All at once there was a quick, high "zinnnng," and I looked around just in time to see Philip yanked off his feet and thrown end over end. Father and I went running to him as fast as we could go, and I could see blood on his neck and the side of his face. Father's hands were shaking nearly as hard as mine when he picked Philip up, but the wire hadn't really hurt him very much. The barbs had ripped the collar off his coat, and had torn a little piece out of the bottom of his ear. It was bleeding all down over his neck. As soon as Father found that Philip wasn't hurt badly, he said to me, "Take him in to Mother. Your punishment will be that you can't ride or drive any horse for a month, and you can't help me with the fence any more."

He didn't say anything about donkeys, but I didn't play with Willie Aldivote's old spotted one for the whole month. Every time I even thought about it, I could hear the "zinnnng" of that wire, and see the red blood the way it looked on Philip's neck.

Father and Mother went to Denver again a couple of days before Christmas. That time they hitched Nig and Fanny to the big wagon and took a whole load of peas. They didn't come home till way after dark. Grace and I could hardly wait for them to get back. She had been telling me that Father and Mother had to help Santa Claus with the Christmas presents, and that they would be bringing them when they came home. We both ran down the wagon road to meet them as soon as we heard the wagon come over the bridge at the gulch. Father stopped the team and let us climb up into the wagon, but there wasn't a thing in it.

While Father was unharnessing, I poked Grace with my elbow and told her she had been making up all that stuff about Father and Mother having to help Santa Claus, but she just looked at me smart and said if they didn't there wouldn't be any presents. When Father hitched Fanny up the next morning and said he was going to the mountains to see a fellow about a dog,

Grace poked me right back and said I'd find out if she wasn't right as soon as he came home. I didn't, though. There wasn't a thing in the buckboard, except his little shingle hatchet, and Grace told me we were too poor for Santa Claus to come that year because the beans got frozen.

Whether Father and Mother helped him or not, we had a fine Christmas. And I never saw anything that looked as though he were getting any help—except the packages that came from our folks back in New England.

Christmas Eve, Mother told us we couldn't get up till daylight, but when the sun first peeked over Loretta Heights we were all dressed and waiting inside the bunkhouse door. Father and Mother were still in bed when we went tearing into the house. There was a big Christmas tree in the corner of their room—all decorated with strings of popcorn and whole cranberries—and there was a big stack of presents under it, but Father said he never even heard the sleigh bells when Santa Claus came.

We all got new shoes and caps with earlaps, and stockings and heavy winter underwear. And I got a jackknife with two blades, and a new geography book. We didn't have any turkey, but Mother baked a whole ham, and we had all the trimmings to go with it . . . and a big plate of fudge.

There wasn't any school between Christmas and New Year's. That's when Fred Aultland started baling his hay. Father and I worked for him all week. Fred said hay had gone to such a low price that he could only afford to pay half as much as he paid us in haying time, but he'd give us ten tons of baler chaff for our week's work. It was good cow feed, and Father said that we could boil it with frozen beans to make the best hog feed in the world.

The baler chaff was all alfalfa leaves and little short stems, so the only way we could haul it was in a wagon box, and Fred said it would take five loads to make a ton. As soon as my month of punishment was over and I could drive horses again, Father let me start hauling the chaff. At first he went with me to be sure I could handle Billy all right, but after that it was my job to go alone and get one load every night after school.

I didn't have a bit of trouble with Billy, but I guess Fanny kind of forgot me during that month. The first day I went to put her bridle on, she kept jerking her head up, so I couldn't get the bit in her mouth. I was standing up in her feed box, and the more she kept bobbing her head the madder I got. At last I grabbed her by one ear to pull her head down. Quick as a wink, she snapped at me with her teeth. She had snapped at me a thousand times before, but had never touched me, so I didn't dodge that time. There was a rip and a burn over my wishbone, and when I looked down blood was coming out of the hole in my jumper.

It scared me a lot more than it hurt, and I went running in to Mother—hollering like a dog with a stepped-on tail. I guess she was as scared as I. Father was working on some little ditch boxes, out in the bunkhouse, and came in to see what had happened. While Mother took off my clothes, he made me tell him what I did to Fanny to make her bite me. Then he just looked at the skinned place before he went back to his work, and said, "Well, I don't see any reason for me to punish you; I think she handled the matter very well herself."

Ever since Christmas, Father had been working on the ditch boxes and a little system of canals. It ran from the well to the far end of the bunkhouse. The Saturday afternoon before Easter, all the ranchers on our side of the creek—clear up to the mountains—came to our place for a meeting. Father explained how the boxes worked so that each one took the right percentage of any water that was coming through the canal. He said that if everybody used them, the ranchers near the head of the ditch would get 60 per cent of the water, and the rest of us would get 40 per cent. Then somebody pumped water into one end of the system, and everybody else watched it work, and said, "Well, I'll be damned."

After they'd played with it till the yard was four inches deep in mud, Father went into the house and brought out a paper he and Mother had been working over every evening for a week. Then he passed it to Mr. Wright, who read it aloud. All the signers agreed to use Father's boxes and not to tamper with them or take any more water than the boxes measured out to them. Fred and Mr. Wright were the last two to go. They both shook hands with Father and told him that if he hadn't figured out the new system, somebody would have been killed in the water fights.

Father told them that he hoped they didn't think his new boxes would be a cure-all. He said that if one man really wanted to be dishonest, another man couldn't keep him from it, but the boxes would make it harder for him to be dishonest without being caught.

Mother and I were proud of him, too. She hugged him around the neck as soon as Fred Aultland was gone, and told him he was the smartest man in the world. I was waiting to tell him the same thing, but he said for me to run along and get the cows.

Our sow—the pig we saved when we were butchering—had her litter on Easter Sunday. There were eight good ones and one runt. Father said the runt would never get his share of milk and would always be sick, so we had a funeral for him in the afternoon. Grace let me be the minister so she could be the head mourner.

From then till plowing time Father was busy every day making ditch boxes. He made them for every rancher between our place and the mountains, but he didn't get any money for them. The men would bring him the boards and spikes and bolts, but none of them had any money, so Father had to trade them his time for little pigs or chickens, or other things we could use. He got a heifer calf from one man and a weanling colt from another. By plowing time we had nineteen little pigs, eight turkeys, and a whole bunch of hens.

It was about that time that I first heard anybody talking about the gold panic. But from then on everybody talked about it. I didn't know what it was, but anyway, you could hardly get money for anything. Fred Aultland couldn't sell his alfalfa, Mrs. Corcoran couldn't sell all her cream, and when Father took peas and beans to Denver he'd come home with more than half of them unsold.

After Father got finished with the ditch boxes, Fred

and Bessie Aultland came to see all the new things Father got for his work. While Mother was telling Bessie where we got this chicken and that turkey, Father and Fred were talking about crops and the panic.

Fred said, "We've all got to face the fact that it's going to be hard to sell anything for money this year. I think you're better off to have got a little stock around you than you would have been to get cash for your boxes. If I was in your place, I'd raise stuff I could feed my stock, and something I could trade in at the store for groceries."

"That's the line I've been thinking along," Father said, "but I don't want to get out more crops than I can get water to raise. Having the stock is fine, but it's left me less than twenty dollars to buy seed with."

Fred always chewed tobacco, and when he was thinking hard he had to spit a couple of times before he said anything. I bet myself he'd spit between the off horse's heels first, but he fooled me—it was between the nigh horse's. Then he said, "I'll tell you what, Charlie, ten of that twenty'll get you seed enough for five acres of sugar beets, you've got beans enough left to sow five acres, and you can flail out oats enough to seed twenty acres from what's left of your last year's crop. I've got a little stack of seed alfalfa that's two years old. I'll trade it to you for four days' work in haying time. With that machine you made, you could clean enough seed to put alfalfa in with your oats, and then you'd have a hay crop all laid down for several years."

Father said, "Fred, you're the best neighbor a man ever had, but I'm afraid you're an optimist. If I should get my full share of water, I'd only have enough for ten acres.

I've already got ten acres of alfalfa; you're talking about my putting down another thirty. I couldn't expect to do much more than lose my seed."

Fred chuckled a little, then he said, "Man alive! You're the only one in the country that will be helped by this damned panic. You don't need money as much as you need food for these kids. I'll make you a bet you'll get all the damned water you need for eighty acres this year. Nobody up the ditch can hire hands this year any more than I can. The big fellows near the head of the ditch can't use all their share of water without help; they'll have to let part of it come on by, except when the ditch is lowest. If you get your ground soaked deep during the spring, and keep a dust mulch over it, you'll have moisture enough to make a crop. By another year your alfalfa roots will be deep enough so they won't need so much summer irrigation."

That's the way we did it. We put Mother's garden and the beets and beans way up at the southwest corner of the ranch—where the irrigation ditch came in—and put alfalfa in with oats on the northwest field. Father got the bean field all plowed, harrowed, and marked off in squares before school let out for the summer. Afterwards, we dropped all the seed by hand, so we would be able to cultivate in all directions and keep a good dust mulch. I would drop five beans where the lines crossed, then Father would hoe some dirt over them and tread on it.

21

I Break Nine Toes

SWEET CLOVER GREW IN THICK ALL OVER OUR LAST year's pea field. Fred told Father it would make pretty good hay if it was cut while it was still young and tender. He let us take his mowing machine to cut it, but Father wouldn't let me go anywhere near the mower while the horses were hitched to it. I'd had my ninth birthday just before Christmas, and had been driving teams for a year. It seemed to me I was old enough to drive the mowing machine just a little while, and I knew it would be fun to sit up on the little iron seat and watch the cutter bar flash back and forth while the clover tumbled down.

I guess I came pretty near begging Father to let me do it, but he said No. Then he told me it was too dangerous, but that he would let me drive the horse rake after the clover had dried into hay.

I could hardly wait for it to get dry enough to rake. I knew just how to kick the foot pedal so the teeth would fly up and dump the hay in straight even rows. I had watched Father do it all one evening the summer before.

When the day for raking came, Father had to put a low seat on the horse rake, because my legs weren't long enough to reach the foot pedal. He used the little iron one off the mowing machine. I could sit clear back in it and still reach the pedal.

At first Billy had been nervous on the mowing machine. The cutter bar went clackety-clackety-clack right behind his heels, and two or three times he acted as though he wanted to run away. But old Nig kept right on plodding, and after three or four times around the field Billy settled down. Father thought he might do the same thing when the horse rake dumped, so he drove the team for the first couple of rounds. Billy behaved as if he'd been pulling horse rakes all his life, so Father boosted me up on the seat and passed me the lines. All he said was, "See if you can keep the windrows straight clear across the field, and don't hurry the team at the corners."

Right at the start, I had a little trouble in kicking the pedal at just the right second to keep the windrows straight, but I got the knack after the first two rounds. Everything went fine till the train came through, and I was planning how I'd be able to get a man's pay in haying time. Old Joe was the engineer on the combination train that went up to Morrison every forenoon and came back every evening. I had known him ever since Bill and Nig fell through the trestle, and we always waved at each other.

I was so busy watching to see that I would kick the pedal just at the right moment—and maybe thinking about being old enough to earn a man's pay—that I didn't even wave to Joe when the train went through. I guess he wanted me to see him wave at me, though. He blew three sharp blasts on the whistle when he was right even with me.

You'd think the whistle might have scared Billy, but it was old Nig that started to run first. He jumped quicker than our tomcat did the time I hit him with a tomato. That made the singletree bump Billy on the hocks, and he took off like a greyhound.

The ground was bumpy where the bean rows had been, and the big high wheels of the horse rake bounced over them so that the iron seat jumped in every direction. The iron was smooth and slippery, and my bottom hopped around on it like a drop of cold water on a hot stove. I couldn't grab hold of anything with my hands because I had to haul on the lines for all I was worth.

Billy could run so much faster than old Nig that we kept turning a little and a little, till we were headed right into the passenger car at the end of the train. It just got out of our way before the horses galloped up over the track. I didn't know I was doing it, but I guess I grabbed hold with my toes when I couldn't hang on with my hands. When the wheels hit the first rail it must have jarred the foot pedal down. The rake teeth flew up to dump the hay. That turned the angle iron bar I was holding on to with my toes, and jammed them in between it and the stay-rod.

When we hit the other rail the teeth flew down again and caught the rail. As the teeth went down, it let my toes loose, but the doubletree broke right at the middle. Of course, that left the team free from the horse rake—and me, too. I had the lines wrapped good and tight around my hands, and I don't think I could have let go if I'd tried. I didn't try; I was too scared to think about my hands.

With the doubletree broken in two, there was nothing to hold the singletrees up, and they kept bumping the horses on the heels. I was skidding along on my stomach with the singletrees jumping around right in front of my nose, and Billy kicking past my head every time his heels got bumped. They dragged me about halfway to the barn before they stopped.

We had been stopped hardly long enough for me to know where I was, before Father picked me up. I didn't know how much I was banged up till then, and I really didn't hurt much anywhere but in my toes. I must have been a little loco, because I don't remember unhooking Billy's trace chains from the singletree, but Father told Mother that's what I was doing when he got there.

They carried me into the house and put me on their bed. I tried not to cry, but I did just a little. It wasn't because I hurt so much, either. It was just because I couldn't help it. And maybe just a little bit because I was glad I didn't get killed.

About all there was left of my blouse was the collar band, and both legs got ripped off my overalls. Mother had hardly taken off what was left of them when old Joe and the train conductor, Mr. Duffy, came to the door. While Father went to let them in, Mother was feeling me all over. Her hands were shaky, and she cried more than I did. The first thing she said when they came in was, "Nine broken toes, and four of them nearly torn off. It will be a wonder if he ever walks again."

Old Joe yanked his cap back onto his head, and started right out again. "Come on, Duffy," he hollered. "We'll highball for the Fort and send Doc Stone out."

Mother wrapped a quilt around me, and Father held me on his lap, with my feet soaking in a bucket of warm water, till Doctor Stone got there. While the doctor was thumping and poking me, and listening to my insides with his little ear trumpet, he had me tell him what happened. After we got done, he looked around at Mother, and said, "You don't ever need to worry about this boy getting killed in an accident; he must have an 'in' with the Almighty. If this leaky heart holds out he ought to live to be a hundred." I could have loved him for that, because the thing I was most afraid of was that Father and Mother wouldn't let me handle horses any more.

After he'd wiggled my toes around some, he told Father to get a piece of smooth board and cut out pieces to fit the bottoms of my feet. When they were ready, he had Father saw slots between the places for my toes.

It hurt like sixty, and I yelled plenty when Doctor Stone was pulling my toes out so as to make the ends of the bones fit together, and while he was taping them down to the boards. He'd put a little wad of cotton under one of them, then wind sticky tape around both it and the place Father had cut out in the board to match it.

He let me rest for a little while after he had set all five toes on my left foot. Then he started to laugh when he was looking at the big toe on my right foot—the only one that didn't get broken. I had a stone bruise on the bottom of it, and he said, "Don't you ever holler about a stone bruise again. If this one hadn't been so sore that your nerves told your brain to keep it up out of the way, you'd have broken all your toes."

I liked Doctor Stone, even if he did hurt me when he was setting my toes. Mother asked him how long I'd have to stay in bed. First he looked at me, then he looked at his watch, and said, "Oh, I'd say till about seven o'clock tomorrow morning. I'll tape these wooden shoes around his ankles, so they won't flop, and it won't hurt his toes any to clump around on them."

My toes and the places where I got skinned up hurt a lot more that night than they did right after I hurt them. Father slept out in the bunkhouse, and I stayed in bed with Mother, but I didn't sleep very much. Before Father went out to bed, he fixed me some brandy in a glass with sugar and water. I got kind of dizzy after I drank it, and I guess I slept some, but it was an awfully long night.

The next morning, Father made me a pair of crutches out of two old broom handles, and I went out to the kitchen for breakfast. Mother had made the other youngsters stay outdoors after I got hurt and when the doctor was there, so I hadn't seen any of them. My toes didn't ache so much that morning, and I guess I was a little bit glad I did get in an accident, because all the others kept looking at me as if I were somebody important.

It's funny how word gets around when anybody gets hurt. The day after I broke my toes, most everyone in the neighborhood came to see me. Even Mrs. Corcoran came and told me I was a fool because I didn't let go of the reins instead of getting dragged. Willie Aldivote brought me a pair of doves that were just big enough to have feathers on them, and Fred Aultland said he knew I was going to make a horseman the minute he laid eyes on me. The one I liked best to have come to see me was King. He acted more sorry than anybody but Father and Mother, and he'd sit beside me for an hour at a time, and every little while he'd lick my hand.

Two Dog and Mr. Thompson came the second day. Two Dog had a little pouchful of dried leaves, and Mr.

188

Thompson told Mother to boil them and put the broth on places where I was skinned. I don't think Mother would have done it if Mr. Thompson hadn't stayed all afternoon to watch. Anyway, she only put it on my hands and arms—and they were the first sores to heal.

Mother let me eat supper out on the porch with Two Dog. He ate all his salt pork and johnnycake, but he didn't touch his beans, and I got Grace to bring out a bowl of sugar to go with his tea. Once in a while he would reach over and lay his hand on the part of my leg that wasn't skinned, and I hoped he'd stay till late in the evening, eating sugar out of his hand, but Mr. Thompson harnessed the buckskins right after supper.

I liked to have people come to see me and ask me about getting hurt. Really, my toes didn't ache very much after the first two days, but I thought it might be nice to act as though they were killing me, so Mother would give me lots of attention and more people would come to see how I was. I tried it for a while the next morning, but it wasn't any fun lying on the bed when Mother was busy in the kitchen and all the other youngsters were outdoors. I couldn't even fool King, and he would only come to the door and whine. By nine o'clock I took my crutches and went out to see how our new colt was getting along. I forgot all about the colt, though, because Father was just coming out of the barn, and called me to come and see Brindle's new calf.

In a few days I got tired of my crutches and threw them away. Father glued pieces of leather on the bottoms of my wooden shoes, so I wouldn't wear out the binding tapes, and I could clump around pretty well. Of course,

I had to walk kind of stiff-legged, the way you do on stilts, but it didn't bother a bit about riding Fanny. Father let me ride her to Fort Logan to see Doctor Stone, so I got a chance to let all the kids in Logan Town see that I had really broken nine toes at one time. All the doctor did was to wiggle my feet around a little and put some fresh bandages on the skinned places. He looked real carefully at the places where Mother put on the broth from Two Dog's leaves. Then he said, "Hmmm, hmmm, I do declare! Find out from that old Injun what kind of leaves those were, will you?" I said I would, but I always forgot it when I saw Two Dog.

That spring, Mr. Welborn, a wealthy man from Denver, had bought the quarter section where I used to herd Mrs. Corcoran's cows. He had an artesian well sunk, and had trees set out along his driveways and where he was going to build his house. He used to pay me fifty cents a day to hoe and water them when we weren't busy haying. My broken toes cost me two whole weeks working for him, but Fred Aultland said I would be worth just as much in haying as if my feet were all right. He said I wouldn't be able to break any more toes driving a horse rake now that I had boards on the bottoms of my feet, and he couldn't see any reason why I shouldn't do it. Fred must have talked to Father and Mother a lot, because they didn't say no.

We did a lot of haying that summer, because nobody but Mr. Welborn had any money to hire help, and the neighbors had to trade work back and forth. My two jobs were driving a hay rake and riding the stacker horse. And, whatever place we worked, Father sharpened the mowing machine knives, fixed the broken machinery, and ran the stacker. When it was our turn, the neighbors all brought their machines and helped us.

I always liked working at Aultland's best. Fred used to butcher a pig for each of his three alfalfa cuttings, so there was plenty of fresh pork, and Mrs. Aultland didn't seem to care how many chickens she fried, or how much sugar it took to make pies and cookies. She and Bessie could cook almost as well as Mother, and they had lots more things to cook.

While we were putting up Fred's first cutting of alfalfa, his cousin came out from Denver for a visit. He brought his wife and Lucy with him. Some of the other men said he was sponging on Fred because he loafed around and told stories a lot of the time. I think his wife and Lucy were sponging, too, because I never saw them help with the cooking or dishwashing, but I liked Lucy just the same. She was a year or two older than I, and while the horses were resting after dinner we used to play up in the hayloft of the barn. She told me lots of things I hadn't thought about before.

Her father had just been fired from a good office job in Denver, but Lucy didn't care. She said he'd been fired lots of times before so it didn't make any difference. I remembered what Fred had told Father about needing food for us youngsters more than money, and I told Lucy about it. Then I said that the Aultlands had better things to eat than anybody else in the neighborhood, and I thought Fred would let them live right there if they did enough work.

Lucy didn't like that at all. She asked me if I thought

her father looked like a darn fool. Then, before I could tell her, she said that only dolts and darn fools lived on ranches, because farmers didn't need any brains and there was too much hard work to do.

When I got mad, she said that Fred and Father weren't fools because they owned their own ranches and hired men to do most of the work. I didn't want to tell her that Father didn't own our ranch, and I didn't want her to think he was a darn fool, so I just kept still. Then she told me that smart men like her father never did have to work hard, because they knew the world owed them a living and there were easier ways to get it than doing hard work.

I wanted her to tell me more about the easier ways, but the men had come out to get the horses, and Jerry Alder yelled, "Jigger, up there in the haymow, Spikes; your old man's coming."

All the men except Father and Fred were there, and when I started coming down the ladder, Jerry called up to me, "I'll bet you learned a hell of a lot of new things up there; did you do any good?"

I told him I didn't know if I did any good, but I sure learned a lot of new things. Then, before I could tell him anything about the world owing everybody an easy living, they all started howling and laughing. Lucy's father laughed louder than anybody else.

While we were milking that night, I told Father what Lucy said about her father, and asked him why he didn't try to do the same thing.

I only saw Father mad two or three times, but that was one of them. He jumped up off his milking stool and came around behind Brindle. His face was gray-white—even his lips were white—and his voice was shaky when he said, "Don't you ever talk to that girl again."

He just stood there for a minute, as if he didn't know what he was going to say, then he put the stool right down in front of me and sat on it. He reached out and took hold of my knee hard. His voice didn't shake then, but he talked low. "Son," he said, "I had hoped you wouldn't run into anything like this till you were older, but maybe it's just as well. There are only two kinds of men in this world: Honest men and dishonest men. There are black men and white men and yellow men and red men, but nothing counts except whether they're honest men or dishonest men.

"Some men work almost entirely with their brains; some almost entirely with their hands; though most of us have to use both. But we all fall into one of the two classes—honest and dishonest.

"Any man who says the world owes him a living is dishonest. The same God that made you and me made this earth. And He planned it so that it would yield every single thing that the people on it need. But He was careful to plan it so that it would only yield up its wealth in exchange for the labor of man. Any man who tries to share in that wealth without contributing the work of his brain or his hands is dishonest.

"Son, this is a long sermon for a boy of your age, but I want so much for you to be an honest man that I had to explain it to you."

I wish I knew how Father was able to say things so as to make you remember every word of it. If I could remember everything the way I remember the things Father told me, maybe I could be as smart a man as he was.

22

Bad Times Were Not So Bad

WE HAD TRADED WHAT WAS LEFT OF OUR GOOD BEANS for groceries at Mr. Green's store in Logan Town, but there were a couple of sacks of our peas and all our landlord's peas and beans left in the bunkhouse. The landlord wrote us in September and wanted Father to bring his share of the good beans in to Denver, and as many of his peas as we could haul.

After the letter came, Mother got down her Wedgwood sugar bowl and poured the money out on the kitchen table. There was only twenty-four dollars and a quarter in it. Most of it was money I got working for Mr. Welborn. Of course, we had receipts for fifteen and a half tons of hay in there, too. That was how we got paid for work we did for other people in haying.

When the money was counted, Mother told Grace to take all the youngsters but me out to play, and to ask Father to come in. The three of us sat around the table where the money and Mother's memo pad were laid out. She said I had earned so much of the money that I should help decide how much of it we could afford to spend and what we would buy with it.

The bottoms of my feet were so tender from having worn my boards all summer that I couldn't go barefoot, and my Christmas shoes were all worn out. Mother said I would have to have new ones before school started, and she thought we'd better pay two dollars and get a good pair, because my Christmas shoes didn't wear very well. Then she said she could can tomatoes and green peas and beans from the garden if we could afford to buy some canning jars and rubbers. After that, we started going over the list of unbleached muslin, calico, quilting cotton, and other things she had written down on her pad.

Really, Father and I didn't do much of the planning. Mother would guess how much different things would cost, and put down the amounts. Then she would add them up, and say, "Oh my! That's a lot more than we can afford to spend. Charlie, do you think there will be a cash market for sugar beets or beans or hay this fall?"

Father only said, "I hope so, but the panic seems to be as tight as it was in the spring; maybe things will open up a little before cold weather."

Mother would cross off some of the things on the list, and change the amounts on others. Then she would sum up, and say, "Oh my!" again. At last it got down to where it was just my shoes, some cloth for winter coats and dresses, and the canning jars—if Father could trade our two sacks of dried peas for them. And then we'd have ten dollars left in the sugar bowl for emergencies.

We loaded the big wagon that night and Father let me go to Denver with him the next morning. We started way before daylight. Mother made me wear Grace's new shoes and the stockings she had knit with yarn from an old shawl. The last thing Father did before we left was to tie the feet of Brindle's calf together and lay him up on top of the load.

Grace's shoes hurt my toes and the stockings made my legs prickle, so Father let me take them off till we were almost in to the Capitol building. First, we went right to the Golden Eagle and got my new shoes; then we drove by Cousin Phil's office, and Father sent me up to get him. The office didn't look a bit the way it used to; there was nothing left in it but two chairs and a desk. The stenographer was gone—desk, typewriter, and all—and Cousin Phil looked terrible. He told me to stay there and look at the newspaper on his desk while he went down and talked to Father.

He was gone a long time, and when he came back he said I was to stay with him while Father went to do his trading. There wasn't a thing for me to do in the office and nobody to talk to. Cousin Phil just kept reading the newspaper and looking at his watch every once in a while. About noontime he went out and brought me back a couple of doughnuts, but he didn't bring himself any and he wasn't gone long enough to have eaten lunch. But he did talk a little after he came back. He told me he had a big deal all ready to go through, but the panic had knocked the bottom out of everything. Then he asked me how I'd like to have Prince to drive instead of old Fanny, and said he thought a while on the ranch would be good for the little bronc.

All afternoon I watched from the window for Father to come back, but it was nearly suppertime before he got there. Prince was tied to the tail gate of the wagon, but Cousin Phil didn't even come to the window to look down. He just told me to run along so as not to keep Father waiting, and said he'd be out to see us before too long.

As soon as I climbed up on the wagon I could see that Father had several good-sized bundles piled up in the front end, and half a dozen boxes with pictures of canning jars on them. When I asked him how he got so many things with only fourteen dollars and a quarter, he kind of chuckled a little bit and pulled his leather pouch out of his pocket. There was still some money in it, and from the sound of the jingle I knew some of the pieces were as big as half dollars. Then he told me that Mother had overlooked the fact that everything was a lot cheaper since money got so scarce, but he wouldn't tell me what was in the packages. He said I'd have to wait till we could all look at them together.

After that I asked him what was the matter with Cousin Phil, and why he didn't have the stenographer and her things in his office any more. He didn't answer me for a few minutes, then he told me we hadn't stopped to realize how well off we were to have enough to eat every day, and to have a good home and clothes enough to keep us warm, when there were people who were actually starving. I remembered about eating both of the doughnuts Cousin Phil brought back at noon, and asked Father if he thought maybe he was starving. He said no, Cousin Phil would never starve, but he thought the panic had clipped his wings a little bit.

Father and I were both as hungry as we could be, and just talking about having plenty to eat made us hungrier. When we got out by the lumber barn we bought a custard pie and a pail of milk in the same little store where we had got them when we were first moving out to the ranch. While Billy and Nig were eating their oats, Father and I ate our pie and talked about all the new things we had been able to get since we had been sitting there eating a custard pie less than two years before. It seemed to us then that we were going to be rich people before very long. And the next morning it seemed to everybody at home that we already were.

School didn't start that fall till we had everything from our garden that we hadn't peddled put away in the cellar. We got some more canning jars and sugar from Mr. Green in trade for a pig and vegetables. Mother filled every one of them with green corn, tomatoes, peas, and green beans. Father built bins on the cellar floor, and we loaded them with carrots, table beets, and turnips; but we had put horse manure under the potatoes when we planted them, and they went all to tops as my cowboy friend, Hi, had said they would.

It was Mother who got the idea about selling vegetables at the Fort. She said that officers in the army got paid cash whether there was a panic or not, and she'd bet their wives would buy fresh vegetables. I think Father would rather have taken a beating than do it, but we went from one back door to another, all through the officers' quarters in Fort Logan, till we'd sold all the garden stuff we didn't need for ourselves. It brought in nearly twenty-one dollars.

Denver must not have been a good place to be during the panic, because Prince looked nearly as bad as Cousin

Phil. He was skinny and didn't have anything like the pepper he used to have when I first saw him. Father started giving him two quarts of oats every morning and night, and by school time he'd run like a rabbit again. Father let us drive him once in a while till he found out what was going on at school, but after that we had to take Fanny. He might not have found out about it at all if I hadn't got my face skinned.

Willie Aldivote rode a horse to school that fall instead of his old spotted donkey. It wasn't much of a horse, but he had a saddle for it—and it was a good saddle. It was wide in the pommel, with a short seat and double cinches. Willie said it was a breaking saddle. We put it on all the horses at school and tried to make them buck, but Prince was the only one that would.

At first he didn't buck very much harder than Willie's donkey used to, but the more we tried to ride him, and the more oats he got, the harder he bucked. After a while he didn't crowhop, but would bounce from one side to the other and turn almost end for end while he was in the air, and he'd get his nose right down between his front hoofs. It got so only Willie and I could stay on him at all—and we got tumbled off plenty of times. I couldn't reach the stirrups, and I didn't dare put my feet into the stirrup straps with my shoes on, for fear one of them would stick and drag me when I got thrown, so I always had to ride barefoot. Even at that, my foot stuck one noontime when I went off sideways, and Prince dragged me a few feet. Miss Wheeler fixed me up all right with a little court plaster, but, of course, Grace had to tell Mother what happened. Maybe it was just as well, though, because after that Prince bucked so crazy there was hardly a man in the neighborhood who could ride him.

Right after Thanksgiving Hal came running up the wagon road to meet us when we were coming home from school. He was just past four years old then, and nearly as wide as he was tall. We could hear him jabbering as he came, and he was so excited he could hardly wait for Fanny to stop before he started trying to climb up over the wheel of the buckboard. In between gasps he was hollering, "We got a new horse . . . and she's got a colt . . . and she's brood . . . and her name is Bread."

I guess we all got about as excited as Hal was, and I turned in at the gate so fast that I nearly slewed him off the seat. Father and Mother were out at the corral with the new mare and colt. The mare was a nice-looking bay, a little bigger than Fanny and as smooth as butter. The colt looked to be a yearling, and was a bright sorrel. Father let me walk right up to the mare and pat her. I wanted to be gentle with her and have her like me, so as I went close to her I said, "Easy, Bread. Easy, Bread."

Mother heard me, and said, "What is that you're saying?" So I told her Hal said that was the mare's name. Mother's face looked kind of funny for a minute, and she pulled her lip down as if she were trying not to laugh. Then she said, "No—no. Hal must have been mistaken when he heard Father talking to Mr. Cash about her. No, she doesn't have a name yet, but she's going to have another little colt pretty soon. Isn't she beautiful? She's a perfect little lady." Mother didn't know it then, but she'd named our new mare. Nobody ever called her anything but Lady after that. Grace named the colt Babe.

Afterwards Father told me that we had bought Lady and Babe with receipts for eight of the tons of hay we had earned during haying time. He said I had a part interest in her, but I could never ride her hard as I did Fanny, because she was a brood mare.

Christmas that year was even better than the year before. Father bought a whole box of yellow Bellflower apples, and we had our biggest young tom turkey and cranberry sauce, and even celery. We all got new shoes and stockings, and Mother had made new winter coats for Philip and me without our knowing about them. And I got a set of dominoes and a two-line bit for a riding bridle. Muriel's cat thought she ought to give some presents, too, and she must have counted noses, because she had a litter of five kittens Christmas Eve.

We had taken three good cuttings of alfalfa off the field we sowed our first year on the ranch, and the stand on the new field was rank and good. Our beans had ripened before the frosts came, we'd had enough water to fill the oats, and there were tons and tons of sugar beets. The only trouble was that we could sell hardly anything.

Father had Mr. Lewis bring his machine and thresh our oats and beans. It took two days to thresh them, and I stayed home from school to carry water and milk and doughnuts around to the men. When the job was done, Father paid Mr. Lewis in oats and beans, but all the other men owed us hay, so we just changed each man's receipt to show he owed us a little less hay. It was fun to watch the stream of clean white beans come pouring out of the threshing machine, and then think about all the work we had had the winter before to flail, winnow, and sort them.

Our nineteen pigs did far the best of all those in the neighborhood, because they lived on sugar beets from the time we started thinning out the rows in the spring. And the sugar beets made the meat so much sweeter that we could trade our pork in at Mr. Green's store when other people couldn't. Grace and Philip couldn't get jobs away from home like Father and me, but they were the ones who really got us the groceries that winter, because they thinned the beets and fattened the pigs.

And it was getting the pigs fat that helped us more than anything else. Everybody had wanted some of our beets to fatten their hogs, and nobody had any money, so Father traded beets for all the tools and machines we needed to make us good ranchers who wouldn't have to borrow. Of course, none of the things he got were new, but Father didn't need new things; he could fix up the old ones so they would work just as well. Our landlord came out just before Christmas, and Father gave him part of our share of the oats and beans for his share of the beets. Father made a deal with him for us to work out the land taxes by helping build roads, too. That way we didn't have to give up so many beans.

Road work started right after hay baling, and I liked it best of all the work we did. The county allowed a dollar and a half a day for a man, and a dollar a day for a wagon and team of horses. Lady had foaled her colt—a pretty sorrel filly that we named Bonny—so she was able to work. And with the wagon and harness Father got in trade for beets, we had two teams to put on road work.

Father drove Lady and Fanny, and I drove Billy and Nig—and they allowed a dollar and a half a day for me just like a man.

All the land was adobe. In the summer the roads baked as hard as brick, but when they were wet the mud clung to wagon wheels till they were a foot thick. The only way to fix that kind of road was to spread gravel over the adobe during the winter and let it work down with the spring rains. The men who had big teams worked on the grader that cut the side ditches and rounded dirt up to form the roadbeds, but those with lighter teams hauled gravel.

Father and I hauled gravel. There was a crew of men, down on the gravel bar in Bear Creek, to do the loading, and another crew did the unloading, so all Father and I had to do was drive team. The bar was on the far side of the creek from the road we were building, and we had to come through the ford with the loaded wagons. Several of the men got stuck coming through the ford, and I was always afraid it would happen to me.

One day Father and I got loaded at the same time, and he was right behind me when I started through the ford. If there was ever a time I didn't want to get stuck, it was when Father was right there to see me. Before Billy and Nig had their feet in the water, I started clucking and popped them with the end of the lines. When the line end hit his rump, Billy jumped ahead and nearly threw Nig off balance. I yelled, "Get up, Nig," and swung the end of the line at him. He was wearing an open bridle and he must have seen it coming, because he lunged into his collar so hard he jerked Billy back against the wagon.

Then I guess I lost my head and started snapping the line-end out the way a mad snake does his tongue. The wagon was right in the middle of the ford, where the sand was deepest, when Father called, "Stop!"

I didn't have to say "Whoa" to the team. There was something in Father's voice that they understood as well as I did. He jumped off his wagon, waded right into the creek, and stood beside my front wheel. "If I ever see you abuse a horse again," he said, "I'll put you at a hard job and give you the same treatment. Now pass me those lines!"

What Father said hurt me so bad my throat felt as if I were trying to swallow a baseball, but it didn't scare me. It was his wading into the icy cold water that scared me. Whenever he got cold and wet at the same time, he always took a bad cold and would cough, sometimes, till there was blood on his handkerchief. I passed him the lines, but I was sure we were stuck so hard it would take another team to get us out. Father drew the reins tight, so both horses were even; then he clucked once, and the team set their shoulders and leaned into the collars.

It was beautiful to watch. At first the wagon didn't budge, but it looked as though Father were pushing on those lines instead of pulling, and it almost seemed that I could see his will passing through them to the horses. The muscles bunched out on their thighs until they quivered, and the wagon inched forward. With their feet planted deep in the sand, they kept it moving, moving, until they were stretched out like show horses in a stance. Then their two nigh hoofs moved forward as if they had both been lifted by the same brain. Step by slow step, the wagon moved through the deep sand and up the bank. As soon as we were on level ground Father passed me the lines and waded back to his own team without a word. I always loved him more after he scolded me than I did at any other time.

While Billy and Nig rested and got their wind, I watched Father come through the ford with Lady and Fanny. He had as heavy a load as I did, and my team was once and a half as big and strong as his. I couldn't see how he would ever get through the deep sand. At the brink of the far bank, he stopped them for just a minute. Then he drew up the lines, and said, "Hup!" quick and sharp. The light mares went down the bank with a rush, over the bar, through the ford, and up the bank. I watched their feet and they were in perfect time every step of the way. I got tears in my eyes, and when Father stopped his team to rest I wanted to go back and tell him I was sorry and would never abuse a horse again, but he waved for me to drive on.

I didn't feel a bit good, and as I came up to the grader Fred Aultland asked me what was the matter. I told him I had got stuck in the creek and Father had to wade in to get me out. Fred knew about Father's cough as well as I did, and he was boss of the road gang, so when Father drove up, Fred sent him right home to get on some dry clothes. When he had started, Fred yelled after him, "And tell Mame to give you a big slug of brandy." I don't know whether she did or not, but Father didn't get a cold that time.

That was the best winter we ever had. New Year's Eve,

Mother got out her little red book and figured up all the money we had taken in during 1908. It was only fifty-four dollars and eighty-five cents, but there was never a time when we were hungry, or when we didn't have railroad ties enough to keep our fire going. Our cellar was full of bins and jars of vegetables and barrels of salt pork. Father had built a little smokehouse where we cured the hams and bacon and pork shoulders with corn cobs. So the bunkhouse rafters were hung with all the smoked meat we could eat till summer. And the floor was piled high with sacks of oats and beans. We even had a half bushel of popcorn.

It was a cold winter with only a little snow, and we didn't have much work to do, except to take care of the stock and saw ties for the fire. But the evenings were the best of all. Grace and Muriel and I would do our lessons as soon as we got home from school, so as to have all evening to play. We learned two plays that winter, but Grace and I usually had an argument over which one we'd do. She liked *The Merchant of Venice* best, because she was Portia; but I liked *Julius Caesar* best, because I was Julius and got killed at the Capitol. Mother had to take most of the long parts like Cassius, but Father was Mark Antony, and even Hal learned the lines for Metellus Cimber. If we weren't doing a play, Mother had us make cross-stitch chair covers while she read to us and Father popped corn or mended a harness.

Some evenings Carl Henry would bring Miss Wheeler over to play whist with Father and Mother. Three or four times, his friend, Doctor Browne, came with them. Those nights Mother let Grace and me sit up till nine o'clock, and Doctor Browne would play casino with us. He liked it better than whist, and we liked him a lot.

23

TORNADO AND CLOUDBURST

THE WEATHER WAS WHAT FRED AULTLAND CALLED "spotty" during the winter and spring of 1908-9. Where we lived, and through the mountains right west of us—up toward Evergreen in the Bear Creek watershed—there was only a little snow all winter, but just south of there, along the headwaters of the Platte River, the snowfall was heavy. That was why Bear Creek ran low all summer, while the Platte and Turkey Creek ran full.

In March we had a tornado. It came on a warm Sunday afternoon when Father and I were down by the creek. It looked like a big black balloon with its tail tied to the top of Mount Morrison. Father saw it first, and called, "Tornado!" Then he started running back toward the house. I ran after him as fast as I could go, but his legs were so much longer than mine that he beat me by five rods. That was the first time I noticed how much better he had got during the winter—he didn't even cough with all that running.

As he went past the house, he called, "Tornado!" to Mother, and kept right on to the barn. We turned all the stock loose and drove it out through the gate—past our barbed-wire fence. Then we propped poles against the house, tipped over the hayrack and wagons, and ran for the cellar. The first hard blast of wind hit us just as we got to the door, but in five minutes it was all over. When we came out, Father showed me where the twister had veered off to the north and cut a regular road up over Larson's hill—two miles straight across Bear Creek Valley from our place.

The second Saturday after that, Mother sent me to take a dress pattern over to Mrs. Larson. I rode Fanny, and went around by the West Denver road and through the ford. I should have come right home, but we didn't have much work to do that day, so I went to look along the tornado path. Father had told me some of the curious things twisters could do, like driving a wheat straw through a fence post without breaking it.

While I was poking around I heard a sound like thunder from way off toward the mountains. I looked up, and there was another big black cloud half hidden behind Mount Morrison. As I watched it, it lifted a little, seemed to draw itself into a tighter ball, and grew a lot blacker. I was sure it must have a tail on it, and that I could have seen it if the mountain hadn't been in the way.

At first I thought about going back to Larson's storm cellar, but I was afraid nobody at home had seen the storm coming, and that it might strike them before they could turn the stock loose and get to the cellar. I flung myself flat on Fanny's neck, slapped her with the line ends, and raced straight for the hillside going down to the creek. There was no time for going around by the ford, and I knew right where to hit the old cattle bridge in Cooly Lundy's pasture below our house.

The hillside was rough pasture land, covered with

sagebrush, Spanish dagger, and cactus; and the rain had washed little gullies all through it. I knew better than to race Fanny down over it, but when I came in sight of the creek I noticed that it had risen nearly to the top of its banks. Then I realized there was a cloudburst coming instead of a tornado, and that the water might be over the bridge before we could get there. Fanny never did like to run downhill, but that day she seemed to know we were racing against the storm, and streaked down across the pasture, dodging sagebrush, leaping gullies, and sliding through shale rock. I had to clamp my knees tight and lie close to her neck to keep from being thrown.

When we hit the edge of the valley floor, Fanny took Lundy's irrigation ditch in a clean low jump, and tore out across the alfalfa field toward the cattle bridge. I don't think I reined her at all; she knew what I wanted to do as well as I did. Thunder seemed to be crashing all around us. I glanced up toward the mountains, but I couldn't see them. The black cloud was lying right against the ground, and between claps of thunder I could hear a roaring up the valley.

The creek was about half a mile from our house, and at that point it clung tight against the foot of a steep, brush-covered hill. A narrow trail led down to the old cattle bridge. As we raced toward it, I could see that the water was clear up to the bridge girders and that part of the bank had washed away, so there was a gap of two or three feet between us and the planking. I was afraid Fanny might not see it, and brought the line end down sharp against her rump to lift her over. While my arm was still in the air, the wind and rain hit us like the cracker of a bull whip.

Fanny jumped and sailed across the open strip of mad, swirling water. It was a long, high leap that carried us nearly to the middle of the span. Her nigh fore hoof thudded down against a half-rotten plank—and crashed through.

I didn't see her fall. I only saw her head go down, and then I was thrown toward the bank like a kitten flung by a dog. I missed the bridge and my arm ripped against the end of the planking as I fell head first into the muddy, rushing water. I couldn't swim, but I don't think it would have made a bit of difference. When my head came up I was five or six feet from the bank. I just caught a glimpse of it as I tried to suck in a mouthful of air, but I got mostly water, and the current rolled me over and took me under again. That time I scraped against a bush, grabbed hold, and hung on. I was lucky, because its roots were in the side of the bank and it swung me around like a picket rope. When my head came up again, I was under a big sage bush that leaned out over the water. I was choking so bad it made my arms weak, and I could hardly pull myself up out of the creek.

I should have thought about Fanny the first thing, but the choking made me sick at my stomach, and for a minute or two I couldn't think at all. Then she squealed. I never knew a sound could hurt like Fanny's squeal. I felt something had hold of me and was tearing me in two. When I hauled myself up through the sagebrush, I could see her muzzle and part of one twisted foreleg sticking above the water that was flowing over the bridge.

I guess I lost my head when I saw it. The current had washed me about sixty feet down the creek, so I scrambled along the bank to the bridge trail and ran toward Fanny. I didn't even stop to think that the water might have carried the planks away, but splashed out onto the bridge. Her leg was broken over sideways just below the knee, and her hoof was caught in the hole it had made. She was straining to hold her head above the water, and her eyes looked up at me as if she were begging me to help her. I don't think I ever planned what I did. I guess I just did it because I loved her. I jumped onto her head and clamped my legs around it. Her muzzle slipped off the end of the planks when I landed, and she struggled once or twice—then my head went under, too.

I only knew the water was turning me over and over. A couple of times my mouth came to the top and I tried to gasp in some air. . . . Then Father's face was right above mine, and his hands were pumping up and down on my back. I was lying with my head downhill, and my face was turned to the side so I could see Father's. It was as gray as ashes.

I couldn't seem to make my body wake up when my head did, and Father's voice sounded a long way off. He kept asking me if I was all right. I was, except that my lungs hurt, but I couldn't say so till after he'd taken me up. He was soaking wet and muddy from the creek, but he opened his shirt and held my chest close against him. As he climbed the trail, I looked back toward where the bridge had been. The water had risen another foot or two, and Fanny's neck and withers showed above it. The rawhide thong Two Dog had braided into her mane floated up and down on the current as though it were waving good-by to me. I think that was what made me cry. I tried to tell Father what had happened, but he had seen the last of it himself. From our house, he had seen me start down across Larson's hill as the storm was gathering, and had run for the creek to warn me back before it rose.

Mother made me stay in bed two or three days till she was sure my lungs were all right, and that my cuts weren't going to get infected. I had a couple of cracked ribs and a little fever, and Father brought Doctor Stone out, but he just gave me some pills and strapped me up. Then he told Mother again that I'd never get killed in an accident. By the time I got up, Father had buried Fanny. I rode Lady down there so I could see her grave, but I was glad I didn't see Fanny.

24

I Become a Cow Poke

THE FLOOD WASHED OUT THE DAM AT THE HEAD OF OUR irrigation ditch and tore away most of the gates and locks. Father and all the other ranchers along the ditch had to work for a couple of weeks to fix up the damage. And then there came pretty near being another battle, and everybody had to go to court to find out how much of the cost they had to pay.

We drove Prince to school for a couple of weeks at the beginning of April. That was after my ribs got well enough so I could go back to school, and Father was using Lady for plowing. After that we had to walk, because Cousin Phil came out and got Prince. The gold panic ended about that time. Fred Aultland started hauling hay again, and Mother sold five pounds of butter for cash.

Mr. Welborn sent a man out from Denver to take care of his trees, and I hadn't been able to make a penny all spring. And I didn't know what I was going to do after school let out May first—I almost wished I were going to herd Mrs. Corcoran's cows again. It would be June before Fred's hay was ready to cut, and there wouldn't be much to do at home, except to weed Mother's garden. And, besides, I was lost without Fanny.

I was talking about it one noon at school, and somebody must have told Mr. Cooper. Anyway, he came over to our house that evening and said he'd heard I was hunting work for the summer. Father told him I was always hunting work anywhere except in Mother's garden, but he thought I'd find enough mischief to get into right at home.

Mr. Cooper lived five miles from our place—over west of Littleton, and nearer the mountains. They got their irrigation water from Platte Canyon, and didn't have any ditch fights, so they always had good crops. He had one of the biggest ranches anywhere around, and always hired a dozen or so men in the summer. Before he went home, he told us he would pay me twenty dollars a month, and give me steady work from May first till the end of September. Then he said he didn't have to have an answer for a couple of days, and he'd drop back and see us again.

I wanted to go to work for Mr. Cooper worse than I'd ever wanted anything. I pestered Father and Mother a lot about it. At first Father said I couldn't go because Fred Aultland had given me work for the past two years and depended on me to ride his stacker horse. Grace could ride a stacker horse just as well as I could, and she didn't think it was fair that I got all the money-making jobs while she had to stay home and help Mother. I went to see Fred on my way home from school the next night and talked to him about it. He said he'd give me twenty dollars a month himself, if the last year hadn't been so tough, but if I wanted to take Mr. Cooper's job, Grace could ride old Jeff.

Maybe that had something to do with Father and Mother letting me go. First Mother made Mr. Cooper promise to let me come home every Saturday night, and Father made him say I could sleep in the house instead of out in the bunkhouse with the men.

Mr. Cooper came for me the Sunday night after school closed. Before we got to his place I knew I was going to like working for him as well as I liked working for Fred Aultland, but I didn't begin to realize how much I was going to like it.

The first one I saw when we drove into his place was my old cowboy friend, Hi. He knew me right away. He was standing out by the corral fence with some other cowboys when we drove in, and he yelled, "Hi there, Little Britches! How many toes you broke so far this spring?"

Mr. Cooper told me Hi was his cattle foreman, and was a great booster of mine. Then he said for me not to let Hi spoil me, but I didn't know what he meant.

After Mr. Cooper had taken me to the house, and his wife had shown me where my room was, I went back out to the corral. There were seven or eight other cowboys there with Hi, and they were talking about bringing cattle down from the mountains for sorting and branding. When Hi saw I had come back, he picked me up and set me on the top rail of the corral. Then he wanted me to tell the other fellows about going up to Two Dog's and getting caught in the cloudburst.

I didn't want to talk about killing Fanny in the flood, and I guess Hi saw I was getting a little choked up, so he asked me where I had put my saddle and blanket. Of course, I'd never had a saddle or blanket, but I didn't like to say so, and said I liked to ride bareback better. All the fellows but Hi laughed when I said that, and one of them hollered, "By God, Hi, that'll learn you not to waste a week's time saddle makin'."

Hi looked kind of funny for about a minute, and I guess I looked funnier. Then he started to laugh, too, "Damn you, Little Britches," he said. "You're going to ride that little old saddle I made you or I'll hang it around your skinny neck."

He reached up and hauled me off the rail, and carried me to the bunkhouse under his arm—the way you'd carry a little pig. Before Father had said I had to sleep in the house, Hi had fixed me a bunk right next to his. He had the quilt spread over my saddle, bridle, and blanket. They were the prettiest ones I ever saw, and I had to bite my tongue to keep from squealing.

It was a breaking saddle like Willie Aldivote's, only a

lot better. The pommel was wide and thick; and it flared out a little before it drew in to the horn, so a fellow could lock his legs in under. The horn was only high enough to get a rope around, and had a nice rake forward—the knob must have been covered half an inch thick with leather. There were wide skirts to the stirrup straps, double horsehair cinches, and rawhide latigo thongs front and back. The blanket was a Navajo—brown with bright green zigzag stripes—and the bridle was silver-mounted with a curb rowel bit. I couldn't believe that Hi was giving them to me—that they were really my own.

It was pitch dark before Hi got through showing me my saddle and making me understand that it was mine to keep. Then Mr. Cooper came out to the bunkhouse and told me it was time for me to come and turn in. He said the boys would do me enough damage when we were out working stock, and he was going to see that I got my sleep when I was at the home place.

We ate breakfast in the cook shack, and the cook was a Mexican who could hardly speak a dozen words of English. But he could make good biscuits and flapjacks, and he put lots of onions and pepper in his fried potatoes. I ate so much that it nearly came out of my ears.

At breakfast Mr. Cooper told me that when we were working with the cattle, Juan, the Mexican cook, would be my boss most of the time, because I'd be the water boy, but he'd do the bossing when we were at the home place. Then he said I could loaf around that day and get acquainted while the men were getting ready for the branding.

When we were through eating we all started out to the

corrals. On the way, Hi said the first thing I ought to do was to pick my horse. I don't think Mr. Cooper liked to have him say it, because he said, "Didn't you hear me tell Little Britches I'd do the bossing around the home place? I think his pa and ma had sooner he'd ride Topsy or Eva."

Topsy and Eva were the little seal-brown ponies Mr. Cooper had driven over to get me. First Hi put my saddle on Topsy and let me ride her, and then he put it on Eva. They were both nice gentle little horses, but they didn't have the get-up-and-get to them that Fanny used to have. Maybe it was my new saddle, and maybe it was because I had been used to Fanny, but I didn't like either of them. Hi's blue roan was in the big pole corral with a couple of dozen other horses, and there was another blue in there that looked almost like him. He was a young horse— wide in the chest and narrow in the withers, the way I liked them. He had a fine black head and sturdy legs with cat hams; I couldn't keep my eyes off him.

All morning the men kept busy roping horses out of the big corral, saddling them, and riding them in the breaking corral. Most always they got the horse they were after with the first throw of their ropes, but there wasn't one of them, not even Hi, who could flip a rope like old Two Dog. Some of the horses busted wide open when they got a rider on them, but most of them only crowhopped around for a few seconds before they quieted down. Hi said there were only two or three of them that hadn't been ridden the last spring, but they had gone a little wild during the winter.

I watched and watched, but nobody put a rope on the

blue. I guess Mr. Cooper and Hi knew I was watching him, and knew I didn't like Topsy and Eva too well. While we were eating dinner, Hi said, "For God's sake, Len, why don't you give the kid a shot at him? I seen him ride his old man's seal-brown down back of the schoolhouse, and with a little learning, he'll stick like a louse."

Mr. Cooper didn't even answer, but kept right on eating till somebody else down the table called out, "Aw, for God's sake, Len, give the kid a break!"

Then Mr. Cooper looked up like he was mad, and said, "Look here, you damn fools, who's responsible for this kid, you or me? I promised his ma I wouldn't let nothing happen to him, and I ain't going to let him fork no green colt."

Hi looked as surprised as could be, and said, "Hell, Len, you ain't been hearin' so good. It's a blue colt we're aimin' to see him straddle, not a green one."

That time all the men laughed, except Mr. Cooper. He pounded on the table, and hollered, "I don't give a damn if he's blue or green or yellow. You ain't going to put Little Britches on no wild cayuses while I'm around to give the orders."

Nobody laughed then, but I saw the fellows on the other side of the table looking at each other out of the corners of their eyes. I don't know how Mr. Cooper could have seen it, because he was looking down at his plate, but I guess he must have. Anyway, in about a minute or two, he looked up at Hi, and grinned. "All right, you dirty sons," he said. "I reckoned that was about what you had in the back of your heads. And I guess it would be safer right here where we got a good pole corral, but I want to see you wear that maverick down before you let this little daredevil fork him."

That was the end of dinner. Hi grabbed his hat, let out a whoop, and ran for the corral with the other fellows right behind him. I did stop to say, "Excuse me," before I got up, but I was second to get to the corral. Ted Ebberts started shaking out a rope as he ran toward the corral gate, but Hi called him back. He said to take it easy, because he was going to gentle-break the colt.

I had seen Father gentle-break a couple of horses, and expected to see Hi go at it the same way. But he didn't. Instead of putting his rope on the blue colt, he tossed it on his own blue, led him out of the corral, and saddled him. When he rode back in, he was holding a short loop on the off side of his saddle, not swinging it the way the other fellows did when they were after a horse. The remuda circled the corral, but Hi didn't follow them. He held his blue quiet near the center till they bunched in a corner. Then he moved in toward them at a slow walk. When they broke, his rope flipped out and settled around the roan colt's neck, the way the tongue of a toad flips out at a fly.

I was watching like a hawk, and I never saw him give his horse the least bit of a sign, but as the blue colt raced out of the fence corner, Hi's blue was right beside him. He was snubbed to the saddle horn with no more than four feet of rope, but there was no jerk on his neck as Hi drew him away from the remuda and into the center of the corral.

For just one second the colt stood trembling. Then he

seemed to explode, striking at the taut rope with his fore hoofs, and thrashing his head to try to shake it loose. My fingernails were digging into one of the corral poles, and I was shaking all over, but Hi seemed as calm as if he had a kitten on a string. His blue circled and moved away, keeping the line snug on the colt's neck, while Hi easy-talked him.

Hi must have held the blue colt there in the middle of the corral for ten minutes. He kept talking to him all the time, as his own blue danced in a circle with the colt thrashing around them. I couldn't make out a word Hi was saying, but his voice sounded like water running over stones in a brook.

The roan was wringing wet, but he had stopped striking. Hi motioned with his hand, and Ted Ebberts swung the gate to the breaking corral open, then stepped away from it. Hi's blue changed the direction of his dance until the colt had been led through the gate without seeming to realize it.

When Ted closed the gate and started toward them with a saddle in his hand, the colt went crazy all over again. Mr. Cooper was standing down the fence a ways from where I was. After the colt quieted down a little, he called to Hi, "Ain't you seen enough yet to know that maverick will never be a kid's horse?"

Hi didn't lift his voice a bit, but he said, "No, I ain't, and you ain't seen the kid ride. You got two surprises comin'."

I liked to hear Hi say that, and I made up my mind that I was going to ride that blue roan if it killed me—but I was really awful scared. I held my hands tight on the rail, so nobody could see how they were shaking, and tried to think of things Father had told me that might make them be still.

As they had moved into the breaking corral from the big one, Hi had shortened the snub rope till the colt's head was pulled to within less than a foot of his saddle horn. He slid from the far side of his horse, dropped his reins, and came around where Ted was waiting with the saddle. His roan stopped dancing the minute he dropped the reins and stood as still as a snubbing post. The blue colt crowded against him, and stood trembling as Hi came slowly toward them with the saddle held chest-high. He was still talking like water running over stones when he eased it over the colt's back.

Hi worked his hand up along the blue colt's neck as Ted moved in closer and passed him a hackamore. He slipped it around the colt's neck and over his nose as Ted tightened the saddle cinch and knotted it. White was showing around the blue's eyes, and every muscle under his dripping hide was pulled tight. He looked as though he might explode any second, but both men worked without a quick move anywhere. My chest hurt and I realized I was holding my breath.

Ted loosened the lasso as Hi passed the hackamore rope through the loop, bent, and took off his spurs. Then Hi hitched up his belt, wrapped the hackamore rope around his hand, and eased into the saddle. When he nodded, Ted slipped off the lariat and jumped clear.

The roan colt stood for maybe ten seconds as though he were cut out of stone, and Hi sat just as still. Then the colt shot off as if a trigger had been pulled somewhere inside of him. I had thought Prince could buck, and that I could ride a bronco, but it was only because I didn't know any better. The blue didn't buck straight out, and he didn't spin or circle. His first leap took his front hoofs ten feet off the ground and they came down like pile drivers. He bounced to the right, smashed down, snapped to the left, and went up again like a geyser. His hindquarters didn't follow his fores, but snaked around like a bucking bull's.

All the blood seemed to have drained out of me and left me dry as prairie dust. My eyes burned and my tongue stuck to the top of my mouth. The roan crashed against the poles at the far end of the corral in a sideswipe, pivoted, and rushed across the ring. He had changed his stride to a chop, and Hi's head was snapping like a ball on a string. It wasn't till then that I noticed he wasn't raking or fanning the roan; just pulling against the horse's bogged head with the hackamore rope, and holding himself tight up against the pommel. The colt was plunging right toward me. Hi saw me just when I was ready to jump, and waved his free arm as the blue jackknifed back toward the center of the ring.

Nothing alive could have stood that pace long, so it probably wasn't more than a minute before the roan rocketed, crashed down, and stood trembling. Hi's face and neck were swollen, and so red they looked as though they might break into flames, but he didn't seem a bit afraid. He stroked the colt's neck and talked to him. His voice hadn't changed a mite from the way it sounded before he got on.

Sweat dripped off the roan like rain from the eaves of a house, and his sides pumped in and out like a bellows. I could see the whites showing around his eyes, and it was a look of fright, not of meanness. None of the men on the fence made a sound as the horse seemed to be making up his mind whether to start all over again or to relax. I watched the quiver in his withers grow less and less, and then he moved a foot forward. The saddle squeaked, and he spooked, but he didn't buck. Then he took another step, and another. Neither Hi nor any man on the fence moved as the colt made a nervous circle of the corral.

After a couple of rounds Hi motioned Mr. Cooper to open the outside gate. The colt shot through the opening, around the big corral, and away across a hayfield. Hi had a short hold on the hackamore rope and was holding the blue's head up to keep him out of another buck. It might have been ten minutes before they came back, and it was easy to see there had been an understanding between Hi and the colt. He spooked a little, and shied off from the corral gate, but Hi let him take his time, and he sidled back through.

When the gate was closed Hi slipped out of the saddle and loosened the cinches. He must have had the rest of it all planned out with Ted Ebberts, because Ted went in and took the saddle to the barn. When he came back he was carrying mine, and laid it by the gate. I was still afraid, but nothing like I had been before, and I knew it was my time to ride, so I climbed down and picked up my saddle. Mr. Cooper took it from me and told me to stay outside till it was on. He asked me if I was scared and I lied to him. I said, "No, not a bit," but I was shaking inside. I never knew a horse could buck like that blue, and I knew I'd go off at the first thud if he did it again. I wasn't so much afraid of falling off outside where a horse could run away, but it seemed as though I would surely get trampled if I went off in that little corral.

Hi gentle-talked the roan colt, and stroked his head while Mr. Cooper and Ted cinched my saddle onto him. Then he motioned me to climb up on the fence beside them. He told me the colt would buck again with a new rider, but not so hard—not so hard as he had seen Prince buck with me. He told me not to be afraid, but to keep myself pulled up tight against the pommel with the hack-amore rope, and to keep my eyes on the roan's ears so I'd know which way he was fixing to jump. After that, he got on his own horse and sidled the colt over against the fence. I noticed that all the other fellows had spaced themselves around the corral with their ropes shaken out. It made me feel a lot safer as I eased myself down into my new saddle.

When I was set Hi wheeled his horse away, Ted and Mr. Cooper jumped back, and I was on my own. The colt bogged his head, leaped, and thudded down. From there on I don't know much about it except what they told me afterwards. But I do know that he didn't buck the way he did with Hi, or I'd have gone flying. When it was over, Hi came riding in to take me off, but I didn't want him to. I was so dizzy I could only see a blur, and I couldn't make words come out of my mouth. Maybe it was because I had bitten my tongue, but I don't think so. I think it was because I was still too scared—and too happy because I hadn't fallen off.

Hi knew what I wanted, though. He said, "You're damn right, you're going to get to ride him. Open the gate, Len!" His blue never left my side more than three feet all the way across the alfalfa field, out over a strip of prairie, and back to the corral. On the way back the colt wasn't fighting; I could feel the smooth power of his muscles under the saddle, and I knew he was going to be my horse.

He had bucked harder with me than the fellows expected him to, and I don't know how I stayed on. I guess I was just too scared to fall off. Anyway, Mr. Cooper shook hands with me after Hi lifted me down. He said, "By God, you're going to make a cow poke, Little Britches. As long as you're with me you can call him your own horse." Then he laughed, and said to the other men, "I thought, by God, the kid was going to pull that one-inch hackamore rope in two before the music stopped."

Father never swore, and I know I wouldn't ever have said it out loud, but before I really knew what I was thinking, "By God, I thought so, too," went through my head.

25

A Pretty Strong Current

I SPENT THE REST OF THE AFTERNOON HELPING JUAN and Hi get the chuck wagon ready. It was really more of a blacksmith and harness shop than a chuck wagon.

Juan's kitchen was only a big pantry with doors at the back. It sat on the open tail gate and was stuffed to the roof with flour, slabs of bacon, sugar, coffee, and potatoes. Two big water casks were fastened to the sides of the wagon body, and Juan's pots and pans hung from the chuckbox like warts on a squash.

After all the branding irons, horn saws, spare saddles, and blacksmith tools had been loaded, it was my job to flush out and fill the water casks. I thought they held a thousand gallons apiece before I got them filled.

Mr. Cooper ate dinner in the cook shack with the men, but he ate his supper in the house with Mrs. Cooper and the little girls. I was nearly through with my second piece of pie when a team drove into the yard, and I heard Mr. Cooper come out of the house and call, "Hiya, Fred."

I thought the answer sounded like Fred Aultland's voice, so I finished my pie as quick as I could and went out. He was so busy talking to Mr. Cooper that he didn't notice me till I went up close to the buckboard, and said, "Hello Fred."

Fred spit so quick he hit the nigh horse on the hock, and said, "By dog, Spikes, I didn't hardly know you.

Where the hell did you get that ten-gallon hat?" It was a pretty good light gray hat. Tom Brogan had given it to me after I rode the blue colt. It was a little too big, though, and he had had to roll up some paper and put it inside the sweat band.

Hi was right behind me, and he came over yelling, "Spikes be damned! This here is Little Britches; top-hand cow poke and bronc buster of the Y-B spread. Light down, you lop-eared old son, and get the kinks out of your legs." Then he started telling Fred about my riding the blue colt the first day he'd ever had a man's hand on him, making it sound as if the colt had bucked a lot harder with me than he really had.

I didn't like to just stand there, so I went over and climbed up on the corral fence to look at my colt. He had been running around the corral until he was sweaty, and his coat glistened blue as the sky in the light of the setting sun. I guess I was thinking about that without knowing it. And about Hi, and the way the colt leaped into the air when he started his buck. The name "Sky High" came into my head before I ever knew where it came from.

It was deep twilight before Hi left the buckboard and came over to where I was. The colt spooked as Hi came up to the fence, snorted, and stared toward us with his head held way up and his nostrils flaring. Hi chuckled, "Lots of fight left in the blue devil yet. God! He's goin' to make a horse." We watched him for a while, and he watched us. At last Hi said, "Didn't want to bust him too hard today. Didn't want to bust his spirit." Then, after he'd rolled and lit a cigarette, "Prob'ly shouldn't ought to of put you on him so quick, Little Britches. Your pa wouldn'ta liked it." He took a couple of puffs from the cigarette and blew the smoke up over the top rail. "But, by God, if he's goin' to be your horse, he's got to get used to you from the jump. Ain't no two ways about it."

I guessed that Fred and Mr. Cooper had been telling him he had let me ride the colt before he was broken enough. I didn't want him thinking too much about it, because I was afraid he might not let me do it again. So I told him what I'd named the colt and asked him if he thought it was all right. "Right?" he said. "Fits him like a glove! Tell you what we'll do, by God; we'll call that old cayuse of mine 'Sky Blue,' and make 'em a matched pair."

It seemed like everything around the place started off with "by God." I told myself I wasn't even going to think it, and then I'd be sure I didn't say it sometime when I wasn't thinking.

I went over to talk to Fred Aultland before he went home, and asked him not to tell Mother about my riding Sky High. He didn't say he wouldn't, but he stuck his hand out to me, and I knew he meant he wouldn't tell Father either.

We pulled out for the mountain ranch early the next morning. I had hoped that Hi would saddle Sky High and take most of the buck out of him, as he'd done the day before, so I could ride him up to the mountains. But he didn't. I was just mopping up the last of the syrup on my plate with a piece of hot biscuit when Mr. Cooper stuck his head in the cook-shack door and said, "You'll

228

be riding Topsy, Little Britches." Then, after he'd started away, he stuck his head back in and said, "I'm giving you orders, Hi! Don't you never let Little Britches fork that blue colt till you've got him plumb wore down."

Juan drove a four-mule team on the chuck wagon. Just as we were ready to pull out of the yard, Mr. Cooper told me again that Juan was my boss away from the home ranch, and that I belonged with the chuck wagon. So I pulled Topsy in beside the near wheel mule.

We waited by the gate while the men got the remuda from the corral and hazed it up the wagon road toward the west. Hi was right behind them with Sky High. He had the colt haltered and his head snubbed up close to his saddle horn. As he went past me, he called, "Figure to give this little old cayuse some halter breakin' on the way up." Sky didn't seem to like it a bit, and plunged around to beat the band. But he couldn't do much about it, because Hi's blue just kept jogging along and not paying any attention to him.

Juan followed with the chuck wagon. Until we were out of sight of the house, I rode along beside the mules, but Topsy didn't like the dust that the wagon stirred up. She kept blowing her nose and bobbing her head. Then Juan waved me to go ahead with the men and yelled, "Adelante, adelante, muchacho!" I had picked up enough Spanish from the Mexican section hands to know what that meant, and dug my heels into Topsy's ribs. I never looked back at the chuck wagon until we were in the little green valley between the hogbacks and the mountains.

I had felt kind of bad that I was only going to be water

boy and helper to the cook, but it turned out a lot better than I expected. Juan didn't want help, even if I had known enough to be of any use to him. All he let me do was carry water for the men and bring in bundles of dry scrub oak for the fire.

Juan had a Mexican waterskin that he tied behind the cantle of my saddle. It was a dogskin, and I don't know how in the world they ever got the dog out of it, because there wasn't a break in it anywhere, except at the neck, tail, and feet. It had been tanned and polished until it was as smooth as a lady's glove, and a brownish-yellow color. The legs hung down on each side of the saddle. They were the drinking tubes, and I had to fill it through one of them. To close them tight enough so they wouldn't leak, all I had to do was fold them over and clamp on a split stick, like a clothespin. The breaks at the neck and tail were sealed so they didn't leak, and were hand sewed with double rows of fine cord that Hi said was catgut.

Every morning that first week Hi took the kinks out of Sky High before he went out to work the cattle. And every morning the colt broke wide open for a few seconds, but the white didn't show around his eyes any more, and he didn't tremble. After he had ridden Sky for a couple of miles, we'd change saddles, and Hi would let me ride him awhile, but he always rode his own blue right beside me. The colt always crowhopped a little after I got on, but he never did any hard bucking. Hi let me ride farther each morning. Then Saturday he tied the waterskin on behind my saddle and rode with me all morning while I took water to the men. Sky High didn't like the legs of the skin dangling against him. I could never tell when he was going to spook or crowhop, and had to keep my knees pinched in tight so I didn't get spilled.

By noon my legs were aching to beat the band from keeping them pinched up so tight on the saddle, and I had a lot of sagebrush scratches on them, because I couldn't always make Sky go right where I wanted him to. While we were eating dinner, Hi told me to put my saddle on Topsy and drag in half a dozen bundles of wood to hold Juan over Sunday, and then we'd get away early for the home ranch.

I didn't stop to have supper with the Y-B fellows at the home ranch, but made Topsy canter all the way, so I'd get home before dark.

Father was just coming in from milking when I rode into our yard. Mother came to the kitchen door, and all the youngsters came running out to see me. I hadn't known I was a bit homesick until I got in sight of our house, but when they all came running out to meet me my throat started swelling up, and I forgot all about my saddle and everything else except that I was so glad to be home.

It was a fine evening. Mother popped corn and let all of us but Hal stay up until ten o'clock. I told them all about the mountain ranch and the dogskin water bag and the chuck wagon. But I didn't say anything about Sky High or the bucking.

Father was awfully quiet, even for him, and I could tell he knew I was holding something back. I think I would have told him all about it if we had been somewhere alone, but I couldn't tell him with Mother and the others there. Whenever I wasn't talking I kept feeling guilty, so I told them all about dragging in wood for Juan's fire, and about Hi having his roan trained so he'd handle any kind of a mean animal without any reining. I said Hi was going to teach me how to train a horse that way.

Father just said that would be a good thing to learn, and that a man who could train a horse like Hi's blue roan would be able to teach me lots of worth-while things about forethought and patience as well as horse handling.

Sunday morning I let Grace ride Topsy up to the corner and back on my saddle. Father went along on Lady, because Topsy was a strange horse, and he wouldn't trust Grace alone with her. Grace didn't like to have him go with her. I think she always did wish she had been a boy so she could have been allowed to do the things Father let me do.

We packed a picnic lunch and spent the whole afternoon down by Bear Creek, but we stayed away from the bridge where Fanny got hurt. Mother had a new book they had bought when she and Father went to Denver to hear Mr. William Jennings Bryan make a speech. It was *The Call of the Wild*, and Mother read to us most of the afternoon. I think I liked that book better than any one she'd read. While she was reading, Father and I whittled a sailboat. That is, Father whittled the boat part and I made the masts and split dry Spanish dagger leaves for the sails. Then Father rigged the sails and booms with string he had brought in his pocket. He fixed two long strings to the main boom so we could swing it from one side of the boat to the other as we walked along the bank.

While Mother and the others were getting supper fixed, Father and I sailed the boat down the creek. At a place where the current wasn't too swift, and where there was a pretty good breeze, we sat down on the bank and Father showed me how we could make the boat go either up or down stream by simply changing the angle of the sail. After I had learned how to do it and was moving the strings so to make the boat tack up against the breeze, Father said, "You know, a man's life is a lot like a boat. If he keeps his sail set right it doesn't make too much difference which way the wind blows or which way the current flows. If he knows where he wants to go and keeps his sail trimmed carefully he'll come into the right port. But if he forgets to watch his sail till the current catches him broadside he's pretty apt to smash up on the rocks." After a little while he said, "I have an idea you'll find that the current's a bit strong up at the mountain ranch."

Just then Mother hoo-hooed for us, so we took the boat out of the water and went back up the creek. While we were walking, Father fastened the strings so the sail couldn't move and tied the long cord onto the bowsprit. When we got to where Mother had supper laid out on the bank he gave the boat to Philip.

We left the creek just when the sun started to dip down over the highest mountain peaks, so I could get back to Cooper's before dark. When I went, Father walked out to the gate beside Topsy. He had his hand on my knee and was looking down at the ground, but he said, "Son, I want you to be a man and do the things men do, but I want you to be a good man. I'm not going to worry about you, but don't take foolish risks—and give the man who's paying you a good day's work. So long, partner." Then he waved to me as he closed the gate.

26

TRAINING SKY HIGH

WE STAYED AT THE HOME RANCH THAT NIGHT. HI rapped on my window when it was just light enough so that I could see the outline of the cook shack against the sky. When I got my overalls on and went out to saddle Topsy, he was waiting for me at the corral gate. His blue was already saddled, and a pair of smooth leather chaps was hanging from the saddle horn. They were just my size and had silver disks along the sides of the legs and around the belt. Hi had cut down an old pair of his own to make them for me.

All the way up to the mountain ranch we talked about Sky High. Hi said there wasn't a mean streak in him any-where, and that he had more brains than any other horse in the remuda, except his own blue. Then he told me to watch and I'd see that Sky always followed the same pattern in his bucking, and that he'd let me ride him from scratch just as soon as I had it figured out.

I didn't have to figure it out, though. All I had to do was close my eyes, and I could remember just how he did it. He'd rear high, bounce first left and then right for six jumps, then crowhop for a hundred yards and go into a stiff-legged run. I guess Hi liked it because I already knew. Anyway, he told me I could try it that morning, but to fall loose if I felt myself going. He said I might just as well get started if I was ever going to do it, because Sky might morning-buck all his first season. And it wasn't because he was mean, but just his way of showing how good he felt after a night's rest. He did buck every morning as long as I was there, and always just the same way. After I got used to it I could have ridden him blindfolded.

Hi started teaching me how to train Sky High right from that day. First it was breaking him to the rein, and teaching him to stop with a light pull on the line, then with just lifting them. By the time we went back to the home ranch the next Saturday he would rein either way without any pull, and come from a lope to a walk when I raised the lines with my hand. After that, Hi filed the rowel out of his bit.

I would have liked to ride Sky High home Saturday night, but Hi thought it would be better for me to take Topsy. I guess he thought Mother wouldn't let me come back if she saw the colt put on his morning show, and he was probably right.

I guess I never noticed how good a cook Mother was, or what good times we had at home, until after I went to work at Cooper's. It wasn't that I didn't like the things we had to eat at the mountain ranch, or that I didn't have a good time when I was up there. I did. It was only sometimes at night, after I was in my bedroll, that I'd even think about home. But always when I got to where I could see our house on Saturday nights, I'd be so homesick that I'd make Topsy run as fast as she could go.

That week end we all went for a picnic up in Bear

Creek canyon. It was the first time the other children had been up there, and I think Father had been planning it long before Sunday came. He had traded our old buckboard, and the colt he got for building ditch boxes, for an almost new spring wagon with two leather-covered seats and red-striped wheels. Mother had the lunch basket all packed when we got up Sunday morning, and Father and I did the chores as fast as we could. You could just barely see the tip of the rising sun when we drove out of our yard.

Father could tell every different kind of tree and rock and most of the bushes and flowers. And he didn't just point them out and say, "That's a spruce and that's a fir and that's a jack pine." He'd show us where this one was different from that one, until even Hal could tell them at a glance. Of course, Hal was too little to go on the hike up the canyon with us, or to climb up the side of the mountain, so he had to stay at the wagon and help Mother get lunch ready. But Father took the rest of us way up into a box canyon he had found when he was hauling fence posts. It was just like a big room, built off to one side of the main canyon, and the walls went up almost straight. He could call one of our names, and the mountains would keep calling it back till it sounded as if they were all full of people who knew us. And he found a smoky topaz for Muriel, and a piece of quartz with green agate in it, that he afterwards ground and polished for Grace.

Mother finished reading *The Call of the Wild* to us during the afternoon, and we didn't get home until time to do the milking. Father said I could have taken Topsy along with us. Then I could have saved about five miles by cutting across to Cooper's place from Morrison. I didn't want to do it, though, because I liked to help Father with the milking, and to have him walk out to our gate with me when I went, and say, "So long, partner."

I learned a lot of things during the six weeks we were at the mountain ranch. My real work didn't take more than an hour a day, and I spent all the rest of the time practicing the things Hi showed me. He taught me how to train Sky High until we could ride the two blue roans side by side, and make them do exactly the same things without even straightening a rein. And he taught me to swing a rope till I could spin it in a flat circle I could walk in, or make it dip to catch the leg of a running calf.

After that first week, I was the only one who ever got a leg up on Sky High, and I must have been on him at least twelve hours every day. As soon as I learned to handle a rope well enough so that I could get it on a calf and shake it off again without getting out of the saddle, Hi helped me break the colt for handling a steer. The first thing we did was to pare his forehoofs right down to the quick so they were tender. Then we set shoes on his hind feet. Always before, when I would snub a calf to the saddle horn, Sky would set his forelegs against the lunge, but Hi said that would be bad with steers. He said that in a couple of years the colt would get sprung knees and never be able to take quick turns. Sky didn't do it any more, though, after his front hoofs were trimmed.

I don't know how many steers I roped in the next few days, but there were lots of them. Sky High's forehoofs got tenderer and tenderer, till by the end of the third day he'd sit down—almost like a dog—and dig his shod hind hoofs into the ground the second my rope settled around a steer's neck. Then he'd take the weight off his forelegs till his hoofs would skim along the ground as light as a dragged hat.

He had just two bad faults that bothered me: he wouldn't always keep a tight line, and he wouldn't always keep his head pointed right at the animal. That way, a breachy steer would rush us every once in a while, or nearly tip us over sideways. I asked Hi what I ought to do about it, and he said, "You take Topsy to peddle water with tomorrow, and we'll let an old bull learn that Sky boy a good lesson."

I lay awake in my bedroll for a long while that night, worrying about what kind of lesson Hi was going to let an old bull give Sky. I had seen a bull rip one horse's belly open, and I didn't want anything to happen to my colt. A couple of times I started to ask Hi what he was going to do, but he was a lot like Father in some ways: he liked to show me how to do things, but he didn't like me to ask questions about it beforehand.

I took the waterskin on Topsy the next morning, and the old bull gave Sky High a hard lesson. Hi put a heavy, double-cinched saddle on the colt. Then he had two of the boys help him catch and halter the biggest bull in the valley. They tied a long rope from the bull's halter to the horn of Sky High's saddle, led them out into the middle of the valley, and took the colt's bridle off.

The bull didn't like the idea of being tied away from the herd. He put his tail up and his head down the minute he was loose, and charged off toward the hills. When he hit the end of that rope, he was at right angles with Sky, and they looked like a pair of acrobats I saw at a carnival. The bull turned a somersault and the colt rolled over onto his back with his heels kicking. Sky High was up first, but the bull was up maddest. That time he didn't charge toward the hills, but right toward Sky. The colt dodged clear and the bull went past him. He circled before he got to the end of the line and charged again. Sky High sidestepped out of the way and raked a chunk of hair off the bull's rump with his teeth. By that time there was a loop of rope lying on the ground clear around Sky. When the bull hit the end of it, it knocked all four feet out from under the colt, and tied him up like a calf ready for branding. Every time Sky would try to get up, the bull would yank on the rope and tip him over again.

I had seen all I could stand, and kicked my heels into Topsy's ribs. As she started I dug my free hand into my hip pocket for my knife, but I never got it out. There was a whistle around my head and Hi's rope tied me up like a chicken for roasting. He could have jerked me right out of the saddle, but he didn't. So when the rope tightened around my arms, I pulled Topsy up without meaning to, and Hi slid his blue to a stop beside me. "Lookin' to get yourself killed?" he asked. "What do you think that bull would do when you lit down to cut that rope? Now you hightail on up the canyon and get some water to them boys, and don't come back till dinner. If that colt can't learn to get out of his tangles, he ain't worth savin'."

I went, but I didn't want to, and I chewed my finger-nails clear down to the quick, worrying about Sky High. I thought sure he'd get a broken leg or his insides ripped out, and every time I gave one of the fellows a drink, I asked him if it wasn't pretty near noon. They were rustling the stragglers down from the draws and gulches where the cows hid away with their new calves, and sometimes I'd have to sit there an hour and wait for a driver to come out into the canyon. When Juan blew his old cow horn for dinner I raced back out to the valley as if there were a pack of wolves after me.

I could see Sky High and the bull from the moment I came out of the canyon mouth. They were still in the middle of the valley, and were having a tug-of-war, but Sky's hind feet were planted deep in the sod and the bull couldn't budge him an inch. The line was tight as a stretched elastic band past the side of his head. I kept an eye on them all the time I was eating my beans and bacon. The bull got tired of the tug-of-war business after a while and started circling again, but Sky High backed away and turned so as to keep a tight rope running past his head.

When I was tightening up my cinches after dinner, Hi came over and noticed that the end of one of my fingers was bleeding. He slapped me on the back so hard it made my teeth rattle, and said, "You stop frettin' 'bout that old cayuse or you'll have your fingers et clear down to the knuckles. He ain't nobody's fool, and I'll lay you no bull will ever dump him again as long as he lives. You stick around here and help Juan this afternoon; I'll let Tom peddle water to the boys in the canyon."

I did go and drag in a couple of loads of firewood, but that's about all Juan let me do besides peel the spuds for supper. He didn't have too much to do himself. I was learning to talk enough Spanish so that we could get

240

along pretty well, so a lot of the time we sat in the shade of the chuck wagon and watched Sky High and the bull. They must have gone up and down the length of the valley a dozen times during the afternoon. I don't know whether Sky ever got dumped again during his life, but he didn't during the time I knew him.

The only other lesson that really hurt me was teaching Sky High to stand ground-tied. When a fellow is working with cattle there are lots of times that he needs to tie his horse where there is nothing to tie him to. He has to be able to go away and leave him sometimes for hours, and find him right there when he comes back. Some horses learn to stand ground-tied after they've jerked their mouths a few times by stepping on a hanging rein, but Sky High was too smart for that. If I left him with the rein hanging, he'd hold his head off to the side so as to keep the line out from under his hoofs, and go back to the remuda. It was a nuisance when I was gathering wood, because I always had to find a bush where I could tie him.

One morning Hi told me to catch up another horse for a couple of days, because we were going to "learn" Sky to stand ground-tied. Hi saddled him, and put on a bridle with short reins and a big rowel in the bit. The rowel was so big that the colt could hardly close his mouth without having it cut against his tongue and the roof of his mouth. After that, Hi got a long iron picket pin with an eye-loop at the top. Then we led Sky High up into the canyon, drove the picket pin clear down to the eye, and ground-tied him within twenty feet of the brook.

There was good grass around the picket, but he couldn't

eat it with the rowel bit in his mouth. And every time he tried to take a step forward or back, the bit would cut the rowel into the roof of his mouth or against his tongue. I got mad about that, and told Hi it was a dirty thing to do, and there ought to be some easier way of teaching a horse. He said, "Yep, they's easier ways, and it would be easier for him to forget. The lessons you remember longest are the ones that hurt you the most when you learn 'em. Do you follow what I'm tryin' to tell you?"

I couldn't help thinking about what Father had said—that night out on the chopping block—and I said, "I guess I know what you mean."

We rode back toward the chuck wagon side by side. Hi kept looking down at the horn of his saddle, but he went on talking. "You ain't goin' to like this, because it'll make his mouth bleed, and he'll slobber a bit, but it ain't going to hurt him much more than it hurts you to get a tooth pulled. After a couple or three hours I'll trade that rowel for a straight spade bit that won't cut him, but he's goin' to have to stand there through all of today and tonight without feed or water. That way he'll learn that he can't move for nothin' less than prairie fire when he's ground-tied. And if he's half the horse I think he is, he'll remember it the rest of his life."

Sky High's mouth wasn't sore for more than two or three days after his ground-tying lesson, and from then till haying time Hi let me work with the cattle as soon as I had dragged Juan enough wood for the day. We were herding out on the rolling prairies between the home place and the hogbacks. Juan would move the chuck wagon from place to place with the herds, and I sometimes had to drag the firewood three or four miles from the nearest scrub oak patch. I had to drag two loads a day, so I always brought one the last thing at night, and then got up at dawn to go for the other one. That way, I could spend the rest of the day at the herds with Hi and the men. I always carried my waterskin on the back of my saddle, and by going from one herd to the other, morning and afternoon, I had plenty of time for training Sky High as I worked.

Hi was range boss, so he went from one herd to the other as I did, unless he had to be at the wagon for branding or dehorning, or something like that. Between herds, we always practiced tricks with Sky High and Sky Blue. We got them so they would even do figure eights at a canter without ever losing step, and so we could stop and turn them both right on the same dime. I think my blue roan liked Hi's roan as well as I liked Hi.

The way Hi had me train him for cutting was to pick some quick-moving steer, or a breachy-looking old cow from the middle of every herd, then work that one to the outside without running it, and drive it a quarter of a mile from the herd. They never wanted to go where I wanted them to, and would duck and dodge to get away from me. At first I had all kinds of trouble, because Sky High would get excited when I had to keep turning and twisting him from one side to the other in trying to work some ornery old heifer out of the herd. Sometimes he would get so mad he'd bob his head and rear. Then we'd always lose the cow we were after, and half the time we'd start some of the other cattle running—and Hi didn't like that. By haying time Sky wasn't as good at heading them off as Fanny used to be, but he could almost always tell which one I was after when I picked it, and would work it to the outside of the herd without my having to rein him enough to make him mad.

27

FATHER AND I LEARN TO SHOOT

I USED TO LIKE HAVING PICNICS DOWN BY BEAR CREEK, so all six of the Sundays while I was working up at the mountain ranch, Mother packed a lunch and we spent the day down there. After the first week, I never did get home on Saturday nights till after dark, and I always stayed as long as I could. So it was always dark when I started back to Cooper's. That's why I never noticed our crops till Hi told me about the ditch fight. From the way Father had acted on Sundays I couldn't have guessed anything was wrong, but I should have known Mother wasn't getting so jumpy just worrying about me.

The last Monday morning I went up to the mountain ranch before haying, Hi asked me if Father had got hurt. I said he hadn't, and wanted to know why he asked me that. Then he told me about the fight on the Bear Creek ditch. He said, "I don't like the looks of things over there. Fred Aultland tells me the gang up the ditch has busted out your pa's patent ditch boxes and is hogging all the water, so your places are drier'n a burnt boot. But he says your pa's got a signed paper that's tight enough to haul 'em into court for damages. If I know them dirty sons as well as I think, your old man better start packin' a .45."

I was afraid for Father, and was going to ride right home and tell him what Hi said, but he told me not to. He said I couldn't tell Father anything he didn't already know, and that Father probably wouldn't thank him for having told me. I still thought I ought to go home, and I would have if Hi hadn't said it would scare Mother to death. Then he promised he'd ride to our place with me the next Saturday afternoon and take his own .45 to Father.

I don't think I ever put in a longer week. I couldn't even find any fun in cutting out cattle with Sky High, and on Wednesday Hi told me I'd have to quit being so nervous with the colt or I'd spoil him. I don't know if being nervous had anything to do with it, but that same morning Sky tossed me twenty feet when I was taking out his kinks.

Only Hi and three of the other fellows were going to stay with the cattle through haying, so Juan started for the home ranch with the chuck wagon right after Saturday's dinner. I should have gone with the wagon, but I wanted to stay with Hi so as to be sure he would ride home with me and take his .45 to Father. It must have been four o'clock before we got the three herds thrown together and had cut out all the horses that would be needed at the ranch for haying. Hi had promised that I could ride Sky High home when he went with me, and I was getting fidgety for fear we wouldn't get there till after dark. I had been thinking all week about how nice we were going to look riding into our yard side by side on our two blue horses.

As soon as the remuda was headed toward the home

ranch, Hi yelled to Barney Ortez, "You take over, Barney. Little Britches and me has got business." He let his roan out into a fast run, but I caught up to him before he had gone a couple of hundred yards. Of course, I was a lot lighter than Hi, and my saddle didn't weigh nearly so much as his, but I'd bet Sky High could have outrun his blue and carried the same weight. After that we let them down into a long lope and kept it all the way to the home buildings. We passed the chuck wagon before we were halfway in.

I never did change my clothes so fast, and I don't believe Hi ever did, either. I wasn't in the house more than five or six minutes, but when I came out he was all ready and waiting for me by the bunkhouse door. He had his .45 on and was holding another gun and belt in his hand. The belt was loaded, and was so long he had to wind it around me twice. As he buckled it on, Hi said, "I ain't giving this to you; you ain't old enough to pack one, but if your pa thinks you might, he'll be more willing to borrow it off of you than he would off of me."

The sun was hanging about a foot above the mountain peaks when we got to the far corner of our ranch. As we rode along the west road, between Fred Aultland's place and ours, I almost felt like crying. Fred's alfalfa looked a lot better than ours, because it was older and the roots were deeper, but it was a sick yellow color and not more than six or eight inches high. Our oat field looked like a desert, and the sweet clover had turned brown along the irrigation ditch.

I forgot all about how we were going to look as we rode into the yard, and could only think about Father and how hard he had worked to get our crops in. Of course, I couldn't have helped any if I had been at home all spring, but it seemed as though I had done something wrong to have been away having a good time while the rest of them had to stay home and see our crops burn up.

Hi must have known how bad I was feeling, because when we got to the corner, he yelled, "Yipeeeee," and threw his spurs in against his roan's belly. We went tearing down the last half mile as if we were running away from a prairie fire, and skidded up to the gate in a shower of dust. The folks had seen us and had come out the kitchen door. Mother was keeping Muriel and Philip back so they wouldn't get stepped on and Father was holding Hal in his arms. Hi didn't make any move to open the gate, so I had a chance to let Father see how well I could handle Sky High when I opened and closed it without getting out of the saddle.

Hi took his hat off to Mother with a big sweep just the way he did the first time I rode his blue, and anybody would think he had always known Father. He called, "Hi there, Charlie. This little old kid of yours is gettin' to be quite a cow poke; broke and trained this here colt all by hisself. He fetched me over so we could show him off a little."

Before I could even get in a word, he yelled, "Yipeeee," and spurred his roan again. I pinched my knees in a bit and leaned forward, and both roans took off together. We tore out through the dooryard, circled around the haystack one way, turned on a dime, and came around the other way. Then we made figure eights side by side—both ways around—and a few quick stops and turns. As we came back to the door, we made the roans keep changing lead so they looked as if they were dancing. Hi's roan would do it with nothing but knee pressure, but we had only been practicing three weeks, so I had to keep turning Sky High with the reins.

Father was always quiet and serious. He wasn't ever sour or sulky, but he just never bubbled over or talked loud as lots of men do. I think the nearest I ever heard him come to it was when we brought the roans dancing up to the door. His eyes were shining, and when he called out to us his voice reminded me of the "merry wedding bells," in the piece Mother used to recite. "Nice handling, Son," he called. And then he said to Hi, "I see you're as good at training boys as you are at schooling horses. I'm proud to have him with you, Hi."

I do think Father was proud, but I know I was a lot prouder. And I could tell by the looks of the other youngsters' faces that they were glad I was their brother. Mother always worried for fear I would fall off a horse and get hurt, but that night she was beaming like a sunrise in the spring. She always waited supper for me on Saturday nights, and she told Hi she was sorry she hadn't known he was coming, because she was quite unprepared, but if he could take pot luck, supper would be ready in about fifteen minutes.

Father went to the corral with us when we unsaddled, but he didn't try to help me. It felt as if that gun and cartridge belt weighed a ton, and the top rail of our corral was pretty high for me to toss my saddle over, but I was lucky and it balanced with the horn pointing straight up on the first try. As soon as we had forked some hay to the roans, Hi unbuckled his gun belt and hung it over the corner post of the corral. I had to climb up on the poles to put mine with it. Father hadn't seemed to notice the gun before, but when I climbed up he said, "That's quite a piece of artillery, Son. Do you wear it while you're working?"

I thought about what Hi had said when he buckled it on me, so I said, "No, not yet, because I don't know how to shoot with it, but I might need it for wolves when I go back to the mountain ranch after haying."

I was just getting ready to ask Father if he'd keep it for me, but Hi beat me. He said, "It would be a nuisance to you during haying time, and I won't be around to learn you how to use it; you might hurt somebody. Why don't you leave it here with your pa till you come back to the cattle? We'll take you out after supper and let you find out how much it kicks."

While we were talking, Mother came to the back door and called, "Su-u-up-perrr," so we went over to the pump and got washed up. Maybe Mother wasn't quite prepared, but she had an awfully good supper. We had a whole roasted ham, and the kind of baked beans nobody else could make—golden brown, with thick juice that was as sweet as maple syrup, and she must have opened a jar of everything in the cellar. After we had eaten till we were full clear up to the ears, and there wasn't a buttermilk biscuit left on the plate, Mother opened the oven door. She had made two pies out of gooseberries Grace and I had picked along Bear Creek in the fall. Some of the juice had oozed up through the leaf pattern she always marked on the top of her pies, and just the smell of them made me hungry all over again. Hi said it was the best supper he ever ate.

At Cooper's I got coffee like the men, and I liked it a lot better than I did milk, but I didn't think Mother would have liked it if she had known. I was a little bit afraid Hi might ask why I wasn't taking coffee when Mother got up for the coffeepot and cups, so I hurried to say, "I won't have room for anything to drink tonight with all these good things to eat." Sometimes Hi could catch on as quick as Father. He was sitting next to me, and when I said that he bumped me under the table with his knee.

After Hi had told Mother how good the supper was, he said to Father, "How about it, Charlie—hadn't we better take Little Britches here out behind the barn and learn him how to shoot a six gun? It's better he have a couple of men around the first time he tries it."

Mother kind of gasped, and looked up at Father. I thought she might spoil everything, but Father kept on stirring his coffee and didn't look up at her. Then he said, "That might be a good idea," almost as though he were talking to the cup.

I said, "Please excuse me," and tried to get up from the table as if I weren't in any hurry, but somehow I got tangled up with the long tablecloth and nearly pulled some of the dishes onto the floor. It was just medium twilight when Mother said, "Now do be careful," and we went back out to the corral. Before we took the guns down, Hi sent me to pick up a canful of pebbles. He said to hunt for ones that were about as big as a peach stone. He and Father helped me, and we must have picked up as many as a hundred. Then we took the guns and went out behind the barn.

Hi dumped the can of stones on the ground about thirty feet from the haystack, then he sat the can right at the foot of it. He had me lie down on my stomach beside the pile of stones, and said, "They's two kinds of gun shootin', but one's all you'll need to learn. In the army they learn you to aim a six gun and shoot it with a straight arm. That's all right if you're target shootin', but it ain't no good for cow pokes. When a poke needs a gun he's always got to make a quick shot, like when a mean bull has knocked your pony down, or wolves has jumped a calf. In them chances there ain't no time to take aim, and you got to be able to throw a slug close to where you want it—and quick. Now pick up one of them stones between your thumb and finger, and fire it at the can."

My first stone didn't get as far as the can by ten feet, and the second one nearly went over the top of the haystack. Hi didn't laugh, but there was a kind of chuckle in his voice, and he said, "See, it ain't as easy as it looks. He ain't got the idea of tossin' it with his forearm and wrist, Charlie. See if you can't show him how it's done."

Father lay down beside me and tossed a couple. His were nearer than mine, but he didn't hit the can. "You ain't quite got the knack yet," Hi said. "It's like this— mostly with your wrist." He flopped down with us and snapped out two or three stones. His wrist acted as if it were hung on his arm with a hinge, and he hit the can right in the middle with every stone.

After that Father could come closer to the can, and he hit it a good many times. I hit it once or twice, too, but

Hi seemed to have almost forgotten I was there. When Father had hit the can about four times in a row, Hi passed him the gun I had been wearing, and said, "Now throw a slug at it the same way. You won't have to think nothin' about squeezin' the trigger." Father whanged a hole through the can the first shot.

We were out there till it got almost pitch dark, but Hi didn't seem to want to stop. He said it was the best practice in the world for a man to learn to shoot after dark, because that was usually the time he had to do his shooting. He said he never did get a thieving wolf in daylight, but he'd got a couple of dozen after dark. Father did more shooting than I did because the gun was too heavy for me and hurt my wrist. He got so he could punch a hole in the can four or five times out of ten shots, but the best I ever did was two.

When it was so dark we could only see a shiny place where the can was, Hi had us stand up and toss stones underhand—the way they throw a bowling ball. Then he showed us how to whip a gun out of the holster, and shoot as the muzzle comes up. He could pull the gun out of his holster, shoot, and put it back, almost in the same motion. He hit the can ten times in a row that way, and drove it—a couple of feet at a time—from one end of the haystack to the other. When I tried it, every one of my shots went into the stack, and Father only hit the can once. Hi said nobody could learn to do it well without a lot of practice, but there were times when it was worth all the work it took to learn.

We had used up all the cartridges out of Hi's belt, and nearly half of the ones in mine before we quit. I went to the corral with Hi while Father took the gun into the house. When he was tightening up his cinches, he said, "Your pa is goin' to make a good hand with a six gun. I don't think I'd say nothin' to him about them fellows up the ditch if I was you. I'll put a flea in Fred Aultland's ear on my way back to the home place."

While we had been out shooting, Mother had made a big pan of fudge. When Hi reined up at the door to tell her again how good supper was, she gave it to him—all packed up in the box Muriel's shoes came in—and asked him if he would mind taking it back to the other fellows at Cooper's. He went off as tickled as I had been with my saddle.

I didn't get any chance to talk to Father about the ditch fight until we were milking Sunday morning. He didn't seem to want to talk much about it then. He just told me not to worry about our crops; that with what we had left over from last year we would be able to get along all right. Then he said there was going to be a court hearing in July, and he thought the neighbors at our end of the ditch were in a good position to collect for the loss of their crops. I asked him if he wasn't afraid there was going to be some shooting that year as there had been the first year we moved there, but he said, "There would be if we tried to take the law into our own hands, but few men will shoot at law-abiding people. If Fred and Mr. Wright can keep the hotheads quiet, I think everything will be all right. Anyway, worrying won't help it a bit, so let's get Mother to fix us up a lunch, and we'll spend the day down by the creek."

We had another fine day down at the creek. I don't

remember what Mother read that day, but I do remember getting her to recite "Thanatopsis," and that she was looking right at me when she said, "So live, that when *thy* summons comes." She said each word slow and clear, and the *"thy"* rung like a stroke on a heavy bell.

I didn't go back to Cooper's that night till after milking. If I had something I needed to talk to Father about, and didn't want anybody else to know, milking time was when we always talked. Most of the nights, though, we didn't talk at all. We didn't that night. I don't know how to tell it, but there was something nice about being out there alone with him and smelling the cow smell, and hearing the milk go singing into the buckets. Sometimes it's nicer not to talk when you're near somebody you love.

Father helped me saddle Sky High when I had to go. He didn't do it as if I were a little boy and had to be helped, but the same way he would have done it with Fred Aultland or any other man. I had never waited till so late in the day to get on Sky for the first time, and I didn't know if he'd buck or not, so I told Father not to be afraid if he was a little frisky, because he wasn't mean. I couldn't get my foot into the stirrup from the ground, but Hi had taught me how to hop and catch the saddle horn and stirrup at the same time. Once in a while Sky High started his leap before I got clear up, and then I had to do it all over again, but that night I guess he knew I wanted to show off for Father, and he didn't rear till I was all set.

He only crowhopped a few jumps and then we waited for Father. He walked out to the gate with me as he always did, but the only things he said were that he was proud of the way I had trained Sky High, and it would be best to put lots of cream in my coffee. He waved to me as he closed the gate, and called, "So long, partner."

28

RIDING IN THE ROUNDUP

Riding hay rake and stacker horse were kind of monotonous after being at the mountain ranch with Hi and the cattle. If it hadn't been for the evenings, I don't think I would have liked it at all. Before Hi went back to the mountains, he told me that I would have to ride Sky High every night if I wanted him to remember all the things I had taught him during the spring, and that I ought to keep him in practice on cutting and roping by working on the young stock in the home pasture.

There were about twenty men around the place during haying. Eight or nine of them were cow hands who weren't needed with the stock through the summertime, and the rest were hands that Mr. Cooper had hired in Denver. All the cow hands were getting ready for the Fourth of July roundup at Littleton. They always had roundups at the fair grounds on Fourth of July and Labor Day, and there were prizes for bronco busting, horse racing, trick riding, and roping.

Mr. Cooper liked to have the Y-B fellows win prizes at the roundups, and kept ten or a dozen outlaw horses at the home place so the men could keep in practice. I wanted to try to ride a couple of them—they didn't seem to buck as hard as Prince used to—but Mr. Cooper

wouldn't let me. And every night he had one of the fellows ride with me when I was practicing with Sky High.
Usually it was Tom Brogan. He wasn't very good at
busting, but he could make a rope do more funny tricks
than a monkey on a grapevine. I learned to do some rope
tricks from him, but he couldn't make his old sorrel do
tricks like Hi's blue, and I never could seem to keep Sky
in step with him.

Mr. Cooper had Hi come in from the mountain ranch
about the end of June, so he could get some practice on
the outlaws before the roundup. And from then till the
Fourth, Hi practiced with me three or four hours a day.
Mr. Cooper saw us riding together that first night after Hi
came in from the mountains, and after that he'd send Tom
Brogan to ride the stacker horse at about three o'clock in
the afternoon, and I had all the rest of the day with Hi.

There wasn't any haying on the Fourth of July, and
everybody went to the roundup early in the forenoon. I
rode Sky High, and I didn't have to stay in the grandstand, either, like the other kids. Hi took me right out into
the middle of the race-track oval, where they had the
bronco busting and bulldogging and calf roping.

Hi won first prize in the bronco busting, and Tom
Brogan in calf roping, but it was the trick riding that I
liked best. One girl crawled clear around her horse,
under his belly, and back into the saddle while he was on
the dead run. Another one stood with her feet in loops of
strap on the saddle pommel, and rode all the way around
the race track without losing her balance. And there were
at least a dozen men riders who did all kinds of stunts,
from going under a running horse to sliding off over

one's tail and jumping back into the saddle.

After everything else was over, the man with the megaphone shouted, "The last number on our program will be an exhibition of matched pair riding by Hi Beckman and Little Britches of the Y-B spread. . . . Bring 'em out, Hi!"

Hi hadn't told me we were going to ride. Tom Brogan picked me up by one arm and the hind end of my chaps, and tossed me into my saddle, but my legs were shaking so that I could hardly get my feet in the stirrups. Then somebody opened the gate and let us out onto the racing strip in front of the grandstand.

It was a good thing I'd had those few afternoons to ride with Hi, because I was so mixed up at first that Sky High pretty nearly had to do everything by himself. I didn't help him much till it came to the end, where we went up to the grandstand changing lead, so that it looked as if both horses were dancing. As we went up, Hi said, "Grab your hat off, Little Britches, when you see me grab mine." I did, and the people in the stands yelled louder than they did when Fred Aultland's bays won the trotting race. All I could think of was that I wished Father could have been there to hear it.

All the Y-B fellows went uptown to Monahan's saloon as soon as the roundup was over, and I went with them. Sky High didn't like going up Main Street very well and kept bobbing his head and dancing. But it was the doctor's horseless carriage that really scared him. He crowhopped right up onto the sidewalk in front of Schellenbarger's market, and he was still trembling when I left him ground-tied at the hitch rail by Monahan's.

I was about as nervous as Sky High, because I knew Mother wouldn't want me going into a saloon. Anyway, not unless I had to go in to see the sheriff. Hi sat me up on the middle of the bar, and lots of fellows came and shook hands with me and called me Little Britches and wanted to buy me birch beer and sarsaparilla. But all the time I seemed to be hearing Mother's voice, as it was down there by the creek, when she recited, "So live that when *thy* summons comes."

I had to talk about something to get that out of my head, so I said to Hi, "I'll bet we could do some stunts pretty near as good as those trick riders. I can do that diving trick and come up on my feet, like you did when Mother and I were planting potatoes." Then I told him about practicing it in the sandy spot when I was herding Mrs. Corcoran's cows.

When we got back to the home ranch, everyone wanted to see me do it—even Mr. Cooper—and I almost wished I hadn't said anything about it. With Fanny, I'd always done the diving stunt bareback, and she never spooked or changed direction when I was starting my dive. I knew it would be a lot different to do it from a saddle, and was afraid I might get a foot caught in a stirrup, or that Sky High might spook so that I'd land square on my head. I guess Hi was thinking about the same things. Anyway, he wouldn't let me try it till we went way out into the middle of a plowed field, and then he led Sky High the first couple of times. I hadn't tried the stunt since about a month before we lost Fanny. And Sky didn't run very well in the plowed ground, so I kind of messed up the first couple of tries. After that Hi let me try it alone and it went better.

From there on to the end of August, I don't think I always gave the man who was paying me a good day's work, the way Father told me to. Hi went back to the mountain ranch with the cattle, and we hardly got one cutting of alfalfa put up before another was ready to be started. But a couple of mornings every week, Mr. Cooper would say he had to go up to the mountains to see how the cattle were getting along, and that I could go with him if I wanted to. Of course, I always wanted to. And Hi would spend two or three hours practicing stunts with me.

There were only two of them that were hard to learn, and we practiced them both a dozen times whenever I went to the mountains, and on Saturday afternoons when Hi came in to the home place. For one of them, Hi would have me stand facing him, then he'd take Sky Blue back a hundred yards to give him a good start, and come pounding down past me. As he came, he'd lean over in the saddle and stick one arm out straight. I'd stick my arm out straight, too. If I kicked my off leg up just at the instant our arms met, and if we got a handhold on each other's arms, I'd go flying right up back of his cantle. The trouble was that I had to kick my leg up before our arms really came together. Whenever they missed, or we didn't get a good handhold, I'd turn a somersault without using my arms. My face got skinned up a little at first, but after a few days I'd sail up back of the saddle nearly every try.

The only other hard one was the one where Hi swung me. We practiced that one first with a jockey pole between the two horses' bridles, so they would have to run side by side without any guiding. We tied our lines around the saddle horns, and when the horses were going lickety-cut, I'd put both arms over my head and lean toward Hi. He'd lean toward me with one arm looped up over his head, and we'd get a wrist hold. Then he'd jerk me out of my saddle and swing me over his head so that, at the top of the swing, I was doing a hand-stand at the end of his upstretched arms. I had to bounce and jump when my feet hit the ground on the off side of his horse, so that he could swing me back into my own saddle again.

It wasn't nearly as hard at it was scary, and we only made two or three bobbles before it worked as smooth as a stream of warm milk. One thing that helped was that I weighed only seventy pounds. Of course, the big danger was that if the horses didn't stay side by side, there wouldn't be any saddle there for me to come down into. After the first few days, though, both roans knew the trick just as well as we did, so from there on we practiced without the jockey pole.

At first I didn't want to tell Father anything about our new tricks, or that Hi and I were planning to ride in the Labor Day roundup. I was afraid he might say it was taking unnecessary chances. Every time I thought about it, I'd feel sneaky and remember about the day I stole the chocolate bar, and what he said to me out there by the chopping block. And how much I liked to have him walk out to the gate with me and say, "So long, partner," when I went back to Cooper's Sunday nights. So I told him that first Saturday night after the Fourth of July, before I even got the saddle off Sky High. I didn't tell him just what the tricks were, but I did say that Hi would look out that I didn't get hurt.

Mother didn't want me to ride in the roundup, but I kind of think Father did. He didn't really tell me I could until the last Sunday. And then he didn't really tell me. There was a paper that everybody had to sign before they could go into the contests. It was something about riding at your own risk, and I wasn't old enough to sign it, so Father would have to if I was going to ride.

I didn't say a word about it, and he didn't either, until we were out milking that Sunday night. Then I heard the milk stop singing in his bucket, and he passed the paper to me down under Brindle's belly. All he said was, "You'd better ride on over to Cooper's tonight. Hi may want to get an early start in the morning."

Always, when I went back to Cooper's on Sunday nights, I'd put Sky High in the big corral, and go right on in to bed. There was always a poker game going on in the bunkhouse, and Father didn't like me to hang around out there.

It was just after dark when I got back that night, and there was only a dim light in the bunkhouse. After I'd hung up my saddle and started for the house, Hi called to me from the bunkhouse door, "Come on over here, Little Britches. Bill Engle left a box here for you." Bill Engle was the express driver to Morrison, but I couldn't imagine why he'd have a box for me.

When I got over there, there wasn't any poker game, and all three lanterns were turned down low. I thought Hi was all alone, but when I went through the door, fellows poured out of every bunk, and started yelling, "Surprise, surprise!" Hi grabbed me up in his arms, and Mr. Cooper turned up the lanterns.

There was a big package sitting in the middle of Hi's bunk. It had SEARS ROEBUCK in big printing across one corner, and the lettering on the tag was so big I could see it before Hi got me halfway over there. It said, "Little Britches, ^{c}/$_o$ Y-B Ranch, Littleton, Colorado."

My hands were shaking so I couldn't untie the strings, and Hi had to cut them with his knife. There was everything in that box that I ever hoped to have. And it was all just my size. There were a pair of mountain-goat chaps with long white hair; a ten-gallon, light tan hat; Spanish high-heeled boots with pointed toes; and a peach-colored silk shirt, with a bright red neck scarf. I didn't find the silver spurs till Hi told me to look in the box again. It must have been ten o'clock before the fellows got done making me try my things on, and let me go to bed—and then I couldn't go to sleep for a long time.

Hi took me to Littleton early Labor Day morning, and I wore all my new cowboy clothes. He wanted Sky High to get used to town noises and hearing the band play, so he wouldn't be nervous when it came our turn to ride. Sky spooked and crowhopped a little the first time we rode up Main Street, and it took him quite a while to get used to the band—they hadn't had one the Fourth of July.

By eleven o'clock he had quieted down, so we put both roans in the livery stable while we went up to the hotel and had dinner. Hi had fixed it up with Father to meet us there, and when we came out after eating, our spring wagon was standing right in front of the hotel steps. Mother and all the youngsters were there, and I don't know when I was any more glad to see them all. I had hoped Father would come to see me ride, but I hadn't ever thought he'd bring the whole family clear down there. I knew Mother would have a fit if she ever knew what kind of stunts we were going to do, and I wasn't a bit sure Father would like it either. Of course, if he'd asked me, I'd have told him, but he didn't ask.

/ Each rider or team in the trick-riding contest had to draw a number out of a hat to see when they would get their turn. We drew the highest number so we had to be last. There were some real good tricks—a lot better than the Fourth of July. The nearer it came to our turn, the more nervous I got, and I think I would have chewed all my fingernails off if Hi hadn't been standing right there beside me. I was sure we couldn't win one of the prizes with all those fancy trick riders, and once or twice I almost wished that something would happen so we wouldn't have to ride at all.

Every nerve in me was singing like a telegraph line on a cold night, when the man with the megaphone hollered, "Hi Beckman and Little Britches on Sky Blue and Sky High, representing the Y-B spread." Then we rode out onto the race track.

Hi couldn't help seeing how nervous I was, and the first two or three tricks we did were the easy ones we had shown Father and Mother at home. I don't know when I got over being nervous, but after the easy tricks were finished, I forgot all about the grandstand being there. Trick riding doesn't take nearly as smart a fellow as most people think, but it does take smart horses—and we had them. Sky High and Sky Blue didn't make a misstep anywhere, and everything went as if we had been practicing for years. We saved the dive trick for the end, and when we raced up toward the center of the grandstand, dived, and bowed, it sounded as if all the Indians in the world were practicing war whoops together. I was lucky. I came clear over onto my feet—and my hat stayed on all the way. I swept it off, the way Hi did, when I bowed.

Father must have thought I was going to get hurt, because he had come down from the grandstand, and when I looked around I saw him standing by the track gate. I guess I forgot where I was, and about Hi and Sky High and Sky Blue, because I dropped my reins and went running down to him. I think I expected him to scoop me up in his arms the way he used to when I was only six or seven years old, but he didn't. He just stuck out his hand and shook mine. Then he said, "Better get your horse, partner; I think the judges are going to call you," but his voice had that silver bell sound in it.

I was nearly back to where Hi was bringing the horses, when the man with the megaphone hollered, "First place in the trick-riding contest: Hi Beckman and Little Britches, of the Y-B spread! Hi, bring Little Britches on over here to the judges' stand."

Most of the men in both Arapahoe and Jefferson counties must have come over to the judges' stand while they were giving Hi and me our gold watches, and I shook so many hands that my arm ached. Mother was still wiping her eyes when Father and I went up into the grandstand to show her my watch. I guess it would have been better if I had told her a little more about our new tricks before we did them. Grace said Mother thought I was going to get killed every minute, and was scared nearly out of her wits.

I let all the other youngsters, even Hal, hold my gold watch and listen to it tick. And after the bucking contest was over—and Hi had won another watch—we all went up to the drug store, and Father bought everybody an ice-cream soda. It was the first one I had ever had, and I liked it even better than birch beer or sarsaparilla. Grace and the other youngsters liked theirs, too, and so did Father, but I think Hi would rather have had whiskey.

29

WE FACE IT

THINGS DIDN'T CHANGE MUCH AT COOPER'S DURING the rest of September, but they hadn't been going so well over to our place. Father wouldn't talk much about the court trial, except to say that it would probably be long-drawn-out. But Fred Aultland told me more about it one night when he was over to see Mr. Cooper. He said Father had rigged some sort of a recording gauge at the headgate of our ditch, so they were going to be able to prove in court how much water had been stolen by the water hogs. He said our neighbors were lucky we had moved there, because if it weren't for Father's agreement and gauge, they would never be able to win damages in court for the crops they had lost. He told Mr. Cooper that Father was going to show his gauge readings in court the next day, and that the water hogs were going to be the most surprised men in the world.

I left Cooper's as early as I could the next Saturday night and got home just before sunset. Father and I put Sky High in the corral and fed him. Then we stood out there by the corral gate quite a while and watched him eat. I don't know just how long we were out there, but it must have been ten or fifteen minutes. We didn't talk. We just stood there leaning on the gate and watching Sky eat. Father was different from most people; you didn't have to talk much to visit with him.

After a while, I told him what Fred had said about our neighbors being lucky we had moved there, and asked him to tell me about his recording gauge at the ditch-head. He said it was nothing but an old coal-oil can he had rigged so that the flow of ditch-water past a paddle wheel would make it turn clear around in a week. Then he had rigged a float with a pencil fixed to an arm. As the water rose or fell, the pencil moved up or down on the paper he had wrapped around the can. He said he had shown the readings in court, the jury had been up there to test it, and had found it to be accurate, so he thought our case would turn out all right.

Mother came to the back door and called us to supper just as he finished telling me about the gauge, so we started to go and get washed up. The washpan and a bucket of water were on the back porch. I had dipped up a pan of water, and was just ready to reach for the soap, when we heard what sounded like a couple of gunshots down the road. It wasn't, though. It was a horseless carriage—the first one that had ever come up the wagon road since we had lived there.

We called Mother and the youngsters out on the porch to watch it come. It was a two-seater, black, with a round hood over the engine. After it crossed the bridge at the gulch, it banged a couple of more times as it chugged up the road toward our house. There were two men in the front seat and two more in the back. When it was almost up to the front of the house, I saw one of the men in the back seat lean over, grab up a gun, and swing it toward us.

Father leaped like a horse going into a low buck, and knocked everybody over but me. I guess I just got bewildered and stood there. Not more than a tenth of a second before the first bullet ripped a hole in our bunkhouse, Father grabbed my arm and yanked me down. There were two more shots. The second one couldn't have missed his head an inch.

The carriage didn't stop, but kept right on up the road. Mother fell back inside the kitchen when Father hit her, and all the youngsters except Grace and me were crying, but Father didn't pay any attention to us. He jumped over Mother as she was getting up, and it seemed less than two seconds before I heard him firing from the front of the house. By the time I got around there, there was nothing but a cloud of dust a quarter of a mile up the wagon road, and Father was standing with Hi's empty six gun in his hand.

He reloaded it as he ran to the corral for Lady. I saw he was going after them, so I ran to the front gate. Lady streaked through before I had it more than half open, and I never saw such a look as was on Father's face.

It was getting to be deep twilight, but it was still light enough so I could see the dust cloud turn south along the road between our place and Fred Aultland's. It seemed ages before I saw the other puff of dust that Lady's feet made when she turned the corner.

Father didn't come back for an hour. Mother wouldn't let me take Sky High and go after him, but she was as worried as I was. She hadn't even cried when Father knocked her over, but before he got back she had bitten her underlip till it was bleeding. She let me stay in the house with her, but she didn't light a lamp, and made all the other youngsters go down into the storm cellar.

When Father did get home, he had Fred Aultland and Jerry Alder with him. They didn't come from the west, though, but from the east, and they were wearing their six guns. Father said the automobile had gone clear around our section, and headed north on the West Denver road. He said it went so fast that he doubted if a man on horseback could have kept up with it for a hundred yards and that it was probably hidden away already in some barn in Denver. He told mother that Carl Henry had ridden to Fort Logan for the sheriff, and then he asked her to get his camera out of the trunk.

He had Fred and Jerry take gunpowder out of a dozen or so cartridges while he was cleaning the camera and putting the plate in it. I wanted to go out to the wagon road with them while they took a flashlight picture of the wheel tracks, but Father told me I'd better stay in the house with Mother because her nerves were all jangled up.

The sheriff came and looked at the wheel tracks and at

the holes in our bunkhouse. He knew me right away, and asked if I had got any more pheasants.

We sat down to supper while everybody was there, but the sheriff was the only one who ate much of anything. He said he would come back the next morning and get the camera plate after Father had developed it, but that all automobile tires looked alike so he didn't think there would be a Chinaman's chance of ever tracing it down. Father had already said he had never seen any of the men in the horseless carriage before, but Fred kept asking him if he was sure one of them wasn't this or that rancher from up near the head of our ditch. Of course, everybody was pretty sure that the shooting was because Father had proof in court about the water stealing, but the sheriff said there was nothing we could do unless we could prove it, and we never could.

Haying was over at Cooper's in early September and, until school started at the end of the month, I worked at the mountain ranch with Hi. It was fall branding time, and Hi was too busy to spend much time with me.

I was homesick. Of course, I knew that if somebody was going to shoot at Father again, my being there wouldn't stop him. But I got it in my head so much that I couldn't think about anything else. And two or three times Hi had to scold me a little because I forgot to take water to the fellows up in the canyons.

I had been so busy thinking about riding in the Labor Day roundup that I didn't notice things around our place the way I should have. It wasn't until I came home that middle Saturday night in September that I noticed that

Billy was gone. I might not have even noticed it then if it hadn't been for milking. Lots of fellows don't like to milk, but I always did. It seemed as if milking was the time when Father and I were kind of away by ourselves, and as if he belonged just to me. He always saved milking on Saturday nights till I got home.

Right after supper that night, Father picked up the big bucket—the one he always used for the Holstein—and lit the lantern. When I started to pick up Brindle's bucket, he said, "Grace is curious to know how you tell which calves on the open range should be branded with the Y-B mark. Suppose you tell her while I do the milking; I'll only be a jiffy." Then he put the lantern over his arm and went out.

I knew right then that there was something wrong. So I told Mother I'd have to water Sky High before I left him for the night. It was a story, though, and I never did it. I went right out to the barn where Father was milking. Brindle wasn't there.

Father heard me come in the door. And I guess he knew what I was thinking, as he always did. He had his head against Holstein's side, and he didn't look up, but he said, "Old Holstein's holding up so well this fall that it would be a waste of fodder for us to keep two cows, so I let Mr. Cash have Brindle."

It was then I noticed I was standing right in Billy's stall, and it was dry and clean. I don't believe I even thought, before I said, "Did he take Billy, too?"

Father didn't say anything till he got done stripping Holstein, but the bunches of muscle were working out and in on the side of his jaw. Then he set the bucket over, and turned around on the milking stool so he was looking right at me. "Partner," he said, "we might as well look it right in the face. We're not going to make it here. We haven't enough feed to see two head of stock through the winter, and I haven't had but five days' outside work all summer. The court has only given us damages for ten acres of crops, and that's all we're entitled to, because we have rights to only ten inches of water. It won't amount to much more than you've earned with Mr. Cooper."

I wanted to say something, but I couldn't think of anything to say, so I just stood there. In a minute Father hung the stool up on the peg, and rumpled up my hair. "Don't worry about it, Son. And let's not worry Mother. There's always a living in this world for the fellow who's willing to work for it, and I guess we're willing, aren't we? Let's go in and pop some corn."

Fred Aultland brought me home from Cooper's the last Saturday before school started. He was there at the home place when I came in from the mountain ranch, and waited for me to change my clothes and get my things together.

Fred and Mr. Cooper were talking out by the cook shack while I was getting my things packed. It was hot and the window was open, and Fred was talking so loud I couldn't help hearing him. "Damn bull-headed Yankee," he was saying. "God and everybody knows we'd never got a dime for our crops if he hadn't rigged that water gauge at the ditch head. And there he stands with a hundred and twenty dollars in his hand for a year's work, and too God damn proud to take a bale of hay from a neighbor. What the hell you goin' to do with a man like that?" I knew he was talking about Father, and I knew Father wouldn't like it, so I grabbed up my suitcase and went out, without even saying good-by to Mrs. Cooper.

Father didn't get home that Saturday night till after I did. He was helping a man build a house over west of Denver. From then till Christmas he just came home Saturday nights, and left before daylight Monday mornings. He did stay home a few days in the middle of December, though. Hal got pneumonia on my eleventh birthday, and until Dr. Stone said he would get all right again, Father didn't go back to work.

I never did know who bought Nig or Lady's two-year-old colt, or the wagons and harness. Grace told me who had bought some of our things, but all she knew about the others was that Father had taken them away and hadn't brought them back. I never asked him, because I knew he wouldn't want to talk about it. When the West Denver job was finished, he let me stay home from school one day, and we went down to Fort Logan to settle up the grocery bill with Mr. Greene. It was eighty-six dollars, and Father let me put my last check from Mr. Cooper in on it. Just before Christmas, he got another job. That time it was helping build a big house in Littleton.

It seemed as though our best Christmases were the ones when we were the poorest. Mother had saved a turkey, and we had all the things to go with it. Packages came from our folks back in New England, and Father must have brought the tree with him when he came home on Christmas Eve. Mother had it trimmed with cranberries and popcorn strung together on long strings, and there were half a dozen oranges hanging from the limbs, like colored lanterns. The presents were wrapped in white tissue paper and tucked in under the tree the way they always were. There was one sled with Grace and Muriel's names on it and another for us boys. And everybody got new shoes and stockings.

It snowed all Christmas afternoon and nobody came to call. Mother had made a big plate of fudge and we popped fresh corn and divided the oranges into sections. We had to do it that way because there were only six oranges and there were seven of us. At first Father said for us not to divide them because they always made his teeth sting, but Mother just laughed at him, and we divided them anyway. I didn't see him squinny up his eye when he ate some of the sections, either. Mother got a new book for Christmas called *When Knighthood Was in Flower.* She must have read us a hundred pages of it that afternoon and evening.

30

WE MOVE TO LITTLETON

WE MOVED TO LITTLETON BETWEEN CHRISTMAS AND New Year's. Father and Mother found a seven-room house on the south edge of town, and Fred Aultland helped us move. There was a barn and a chicken house, and a little piece of ground where we could have a garden. Besides King, we took Lady, Babe, and the chickens with us.

We didn't live very far from the schoolhouse, and Mother took us over the first day after New Year's. It seemed to us like an awfully big school; there was a separate room for each grade. After the principal had asked us some questions and had us read to him, he put Grace in the eighth grade and me in the sixth. Muriel went into the fourth grade and Philip in the second.

Starting school in Littleton wasn't a bit like starting in at the ranch. Of course, I didn't know any of the kids, but they all knew who I was. I guess there had been something in the paper about my riding in the roundup.

It was right after we moved to Littleton that Father was made boss on the house-building job. I don't think I ever saw him more pleased about anything. He told us about it one night when we were eating supper. I knew he had been worrying about the house, because I had heard him tell Mother the framing wasn't true and there'd be trouble when they went to put the roof on. That night at supper he told us the owner had come out and caught them splicing rafters that had been cut too short. Mother took a quick little breath, and said, "Charlie, does that mean—"

Father looked up and smiled. "Yes Mame, that means—" he said, "that he made me boss carpenter. I'm getting four dollars a day, and I know I can make a good job of it." He took a couple more mouthfuls and then he looked up again. "How does that line near the end of Hamlet go? The one about there being a divinity."

Mother knew them all, I guess. She got tears in her eyes and in her voice, too. "There's a divinity that shapes our ends, rough-hew them how we will," she repeated.

Father nodded, "That's the one. How do you remember them all, Mame?" I think that pleased her as much as the four dollars a day.

I had gone to school in Littleton about six weeks before I got into any big trouble. The teacher in our room was a widow. She was almost a Mrs. Corcoran kind of woman. I don't think she ever said anything nice if she could find a way to say it mean. The only times she was really pleasant were when Mr. Purdy brought eggs and butter to her. Mr. Purdy was a widower who lived four or five miles up the Platte River, and he used to bring the eggs and butter during school. Sometimes they would stand at the door of the schoolroom for nearly half an hour, whispering and giggling.

Mr. Purdy came to the door one day in February—just after recess—and just after they had put new gravel on the school yard. The yard was wet, and we had all lugged gravel in on the soles of our shoes. When Mr. Purdy had talked with Mrs. Upson for nearly fifteen minutes, one of the boys started to scuff his feet back and forth. Inside of a minute everybody in the room was scuffing, and it sounded like forty steam engines all puffing at once. Mr. Purdy left in a hurry, and Mrs. Upson went flying out after him.

She was back in two minutes with the principal, but the room was as quiet as if it had been empty. The principal was a big, handsome man with wavy brown hair and red cheeks. I don't suppose he was more than a year or so younger than Father—probably thirty-two or three—but he didn't look within ten years of being as old. He stood up in front of the class and clapped his hands, then he said, "I want all the children who scraped their feet to stand up."

Dutch Gunther was the first one up, and his brother, Bill, was right behind him. When I looked around, there were seven of us boys standing—and not a single girl. There must have been thirty of us in the class and, if the principal had bothered to look, he could have seen scratch marks on the floor under every desk. He folded his arms and glared at us for a couple of minutes. Then he said, "I might have known—the worst boys in the whole school! YOU FOLLOW ME!"

He marched out of the room like one of the drill sergeants over at Fort Logan, and we marched after him. When we were going through the coat corridor, Dutch whispered back to me, "Don't let him make you holler, Little Britches."

He led us down to a room in the basement, and took a whip off a hook on the wall. It was a mean-looking whip. It was like a bullwhip, except that it was only about a foot and a half long, and it had three cattails at the end. Bill got fourteen licks before he hollered, and three afterwards. I didn't do so well. I had cracked a couple of ribs at the time we lost Fanny, and knobs had grown over the cracks. The first time he swung the whip, the cattails hit right over the knobs, and it felt as if I were being stabbed by a dozen broken bottles.

I thought Mother would go wild when I got home. She would have gone right over to the schoolhouse if I hadn't told her it would only make it worse for me. She washed the places where the cracker cut through my skin, put some salve on, and put me to bed. Afterwards she brought me up some brandy with sugar and water, but it didn't taste as good as it used to, and my back was so sore I had to lie on my stomach.

She must have told Father as soon as he got home from work. He hadn't been in the house more than a few minutes when I heard him coming up the stairs. After he said, "Hello, Son," he turned down the bedclothes and looked at my back. I couldn't have told by the sound of his voice, or what he said, but I knew he was mad because those muscles at the sides of his jaws were working out and in. After he'd looked at all the welts, he said, "Gave you a good one, didn't he? Well, you've been hurt worse than this and got over it—I guess you'll live. Let's get some clothes on and go down to supper."

While I was dressing, he sat on the edge of my bed, and said, "You know, Son, sometimes a fellow has to take a licking for doing the right thing. A licking only lasts a short while, even if it's a hard one, but failing to do the right thing will often make a mark on a man that will last forever. Let's go down and eat."

Father's house was pretty nearly finished. At supper he said there would only be about another week's work, but a man had come to see him about building another, and he was going to start on it the tenth of March. He talked more at supper than I had heard him for a long time, but he didn't say a word about my getting a whipping at school. Grace started to say something about it, but he kept right on talking about the house, so she had to keep still.

Mother sent us all to bed as soon as the dishes were done, but I couldn't go to sleep. I must have lain there about an hour when I heard Father go out the front door. It was about an hour before I heard him come back.

He called me to get up at the regular time in the morning, and when we were eating breakfast I noticed that his hands were all swollen up and dark-looking across the backs. I wondered what he had been doing, because I was sure I would have noticed if they had been swollen like that when he was talking to me the night before. I thought I could figure it out if I could find out where he had been, so I asked him if I didn't hear him go out somewhere. He was wiping syrup off his plate with a piece of hot biscuit, and said, "Oh, I just had to go see a fellow about a dog."

Mother looked up quickly and said, "I think you got it backwards," but Father just kept wiping up syrup.

Grace had gone back after school and got my coat and cap, and Mother didn't say anything about not going to school, so I went. I think I must have gone past the principal's office seven or eight times that day, but I never saw him. The door of his office was always open but he was never in there. He wasn't there for several more days, either. The kids said somebody had given him an awful beating, but I guess I was the only one who ever had an idea who the "somebody" was. I never even told Grace.

Father finished his house on the fifth of March. I remember the date as well as if it had been yesterday. Ever since we had moved to Littleton, Father had been

278

planning to fix Mother's chicken house, but he was never home in daylight, except on Sundays. The first day after his job was finished he started on our chicken house. I went out to help him as soon as I got home from school.

He must have been thinking about the licking I got from the principal, because I had only been working a little while when he said, "You're getting to be quite a man now, Son. You're well past eleven years old, and you can do quite a few things better than a good many men. I'm going to treat you like a man from now on. I'm never going to spank you again, or scold you for little things, and some day it's going to be 'Moody and Sons, Building Contractors.'"

31

SO LONG, PARTNER

I HAD NEVER KNOWN LADY'S OLDEST COLT MUCH TILL we moved to Littleton, because Father had always pastured her away from our place. After we moved to Littleton he began gentle-breaking her on Sundays. There really wasn't much to it. She was a beautiful thousand-pound sorrel, and as gentle as Lady. By the time Father finished his house-building job he could drive her almost anywhere.

The morning after we fixed the chicken house he was talking about her at breakfast. Lady hadn't had a colt the year before and wasn't going to have one that year. Mother said it was a shame not to be raising a colt after the good price we got for Lady's last one. Father looked up and said, "What would you think about Babe? I've been thinking I might drive her up to Fort Logan this afternoon. Judge Rucker's got a horse up there that I think might make a good husband for her."

I hadn't been home from school more than five minutes that afternoon before Doctor Stone brought Father. They were leading Babe behind the buggy, and there were wire cuts on her shoulder and off foreleg. Father had court plaster on the side of his face, and his arms weren't in the sleeves of his coat. When he got out of the buggy I could see that his leg was bandaged. His overalls were torn half off one leg and the bandage showed through.

Mother, Grace, and I ran out to meet them. We were scared to death, but Father grinned and said it was nothing; that he had just been scratched a little. Doctor Stone didn't talk that way, though. He said it was lucky Father was still alive. After he and Mother had put Father to bed, they came out into the kitchen, and Doctor Stone told us what had really happened.

There were big iron gates at the entrance to Fort Logan, and brick walls ran back both ways. Anyone driving on the road outside the wall couldn't see a team coming out of the Fort till it came through the gates. Father and Babe had been almost up to the entrance when a horseless carriage came racing out of the Fort. Babe had never seen one before and reared. The man who was driving the machine tried to stop it, but it went into a fit of backfiring. Babe whirled off the road and plunged into a gully with a barbed-wire fence running through it. Father was thrown out when the buggy upset, but jumped up and flung his weight onto Babe's head, so as to keep her from destroying herself in the wire. He was badly bruised and torn before he quieted her.

That summer on the ranch, without any crops and only a few days of haying, had been good for Father's lungs. Until the night he was hurt, I don't think I had heard him cough in months, but that night I could hear him long after I had gone to bed. It must have been that he got his chest squeezed when he was wrestling with Babe down there in that gully.

Father called me as usual the next morning, but he looked bad when I came down to breakfast. Where it wasn't skinned, his face was gray, and he had a little hacking cough that sounded as if it started clear in the bottom of his lungs. It was one of those cold drizzly March mornings, and Mother wanted him to go back to bed, but he wouldn't. He said he had promised the undertaker he would dig a grave that day, and it might be his only chance to build a house that would last until doomsday. Mother didn't like it, and said that was no time for banter, because if he worked out in the rain in his condition, he might be digging his own grave. Father chuckled a little when he got up from the table, and he rumpled my hair. "We Moodys are tough fellows, aren't we, Son?" he said. Before he went out, he laid his hand on Mother's shoulder and said, "Don't worry, Mame, I'm not sick; I'm just scratched up a little. This job will only take half a day, and there's three dollars in it."

The job did take longer than half a day. I had been out of school an hour before Father got home. Mother had him put dry clothes on right away, and made him drink some brandy and hot water. I don't know whether it was the brandy that made Father talk that night, or whether he had a premonition. He had never told us youngsters anything about his boyhood, or things he had done before we were old enough to remember. That night we sat at the supper table for nearly two hours while Father told us about the little backwoods farm in Maine where he was brought up by his deaf-mute father and mother. And about going to visit his uncle's family when he was eight years old, so that he could learn to talk with his mouth as well as his fingers. He told us about grafting apple boughs onto birch trees, and about lowering himself down into the well so he could see the stars in the daytime. But he didn't tell us anything about being the New England bicycle-racing champion—Mother told me about that afterwards.

I heard him coughing every time I woke up during that night, and the next morning he stayed in bed. The doctor from Littleton came that evening and said Father had pneumonia. He was so sick that the doctor would only let us go in to see him once during the next week. Mother had sent us all to take a long walk on Sunday afternoon so as to get us out from underfoot. She had spent almost every hour with Father since he was taken sick, and her nerves were so unstrung that we irritated her.

When we came home from our walk, the doctor said we could each go in and see Father for just a minute. Grace went first, and then it was my turn. He looked so bad it frightened me when I went into the room. I couldn't think of a thing to say, and I guess Father was so sick he couldn't either. I had found a coil of inch rope lying beside the road when we had been walking, and had brought it home. I could only think to tell Father about the rope. He raised his hand up a little, and I took it. His voice was almost a whisper, and he said, "You take care of it, partner, you may need it."

That was the last thing I ever heard him say. Afterwards Mother told me he had asked for me his last day, but the doctor wouldn't let her send to school for me.

When we got out of school at noon—ten days after Father was taken sick—Hal was waiting for us with a note. The doctor had sent a nurse to help Mother for the past few days and the note was in her handwriting. It said for us to go to the Roberts' house for our lunch, and not to come home. They lived a block nearer the schoolhouse than we did, and were good neighbors to us. They had the only telephone in the neighborhood and, while we were eating, the nurse came in to use it. I think it was Cousin Phil she called. After she'd told who she was, she said, "We've got to have a tank of oxygen out here right away. Yes. Yes, it's got to get here right away if it's going to do any good."

Hal was waiting for us with another note when school let out. That one named different houses for us to go to until Mother sent for us. I was to go to the Roberts'. When I got there Mrs. Roberts gave me a piece of bread and jam. I was standing just outside the parlor door eating it when the nurse came in. She didn't say anything to Mrs. Roberts or to me, but walked right across the parlor and cranked the telephone. I thought it might be something more about oxygen, so I stepped over where

I could hear better. The nurse spoke a number into the telephone, and in a minute she told who she was and said she was talking for Mother. Then she said, "Her husband died about twenty minutes ago. You better pick the body up right away. I want to get rid of it as soon as we can; her nerves are going all to pieces."

It was too big for me to take all at once like that. I didn't feel like crying—I didn't feel like anything. My brain just stopped working for a minute or two. When it started up again it was going round and round like a stuck gramophone cylinder, and was saying over and over, "So long, partner; so long, partner; so long, partner."

Bessie and Mrs. Aultland came to stay with Mother that night, and we youngsters stayed where the note had told us to. My mind was sort of numb during the days between Father's death and the funeral. Things that happened still seem unreal. I do remember that I got a new blue serge suit—the first suit I'd ever had that Mother didn't make—but I don't remember where it came from.

All our old neighbors from the ranch were at Father's funeral, and I never knew till then how much they really cared for him. After the services, Dr. Browne glanced at Mother's red-streaked hand and said, "Mrs. Moody, that is surgeon's blood-poisoning. If you're ever to raise Charlie's children, you must come home with me at once."

Everybody was shocked except Mother. She was a small woman, and Doctor Browne was a very large man. She looked up into his face and said, "Yes, Doctor, I know. I believe I have no choice in the matter."

All our neighbors, both from the ranch and from Littleton, pressed around, offering to take us youngsters in. Cousin Phil said something about writing our other relatives in New England. For just one moment, Mother's eyes flashed; then she was calm again. "No, Phil, I am sure Charlie wants us all to be together."

Then she parceled us out to near neighbors; being sure that Hal went where there was a good cow, and that Muriel went to a motherly woman without too many youngsters of her own. At the end she said to me, "Son, I want you to stay with Laura Pease, where you will be near home and can take care of Lady and the hens.

"Tomorrow you take Babe over to Mr. Hockaday and tell him Father would have wanted him to have her. He needs a good horse, and he's a fine, honest man. He'll pay us all she's worth."

Then she thanked our neighbors and kissed us all around, leaving me till the last. I remember how my lip trembled, wondering if I were the least. She didn't cry until she put her hand on my head, and said, "You are my man now; I shall depend on you. Mother will be home in two weeks."

It was not two weeks, but four. At the end of the first week, before Doctor Browne was sure he wouldn't have to amputate the arm, Mother sent for Grace and me. Grace had her thirteenth birthday two days after Father died. We harnessed Lady to the spring wagon and drove to Denver, stopping by the river to gather a bouquet of pussy willows.

At Doctor Browne's big house on Capitol Hill we were only allowed to see Mother for a few minutes. She was so thin we hardly knew her. Her eyes were deep in their sockets, with black circles around them; and for the first time I noticed white in her hair. Her voice was very low, almost a whisper. She put her good hand out to us and smiled. "Mother is going to be all right," she said. "I have talked to the Lord a lot about it. He knows you need me, and with Him and Doctor Browne, I shall be all right."

Doctor Browne started to lead us from the room. When we had reached the door, Mother called me back. She took my hand and said, "The peas should have been planted on Saint Patrick's Day. You know where the seeds are in the barn loft. Soak them overnight, and put plenty of hen manure deep in the trench." I don't know why that made me cry when I hadn't before. But from that moment I was sure she was coming home.

It was late in the afternoon of a pleasant mid-April day when they brought Mother home. Cousin Phil drove her out in his first automobile—a two-cylinder Buick with shiny brass rods to support the windshield. Doctor Browne and a nurse came with them. They carried Mother into the house and put her to bed downstairs in the parlor. When I came in she was saying to the nurse, "I am perfectly all right now; all I need is my children." As quickly as I could get out, I harnessed Lady to the spring wagon and started the collection of brothers and sisters.

Mother could be quite persuasive if necessary. She must have been so with Doctor Browne because, just as we turned into the lane, the Buick was pulling away from our house. Doctor Browne and the nurse waved to us from the back seat as they went by.

I was the last one into the house, because I had to unhitch Lady. Most of the tears were shed before I got there, and Mother was propped up in bed with Hal still sobbing and trying to bury his nose in her side. Her right hand was heavily bandaged.

When I came in she organized the first meeting of the clan of Moody. "Now let's not be sorry for ourselves any more," she said; "we've got lots of other things to do. First, we must get Mother's hand well. All it will take is good food and good care. I can't think of anything that would be better for it right now than a good chicken stew.

"Ralph, suppose you dress that big fat Buff Orpington hen that didn't lay last winter. Philip, you get Grace two or three armfuls of wood and some shavings, so she can start a fire in the cookstove. And Muriel, do you think you could get the new tablecloth out of the dresser drawer, and set us a table right here by my bed? When you get the fire going, Grace, put on the big iron pot with some fat in it so it will be good and hot when the hen is ready. And, Hal, would you get Mother a drink of water? I can't think of a thing that would taste so good as a nice cool dipper of water, right from our own well."

That first supper was the most memorable meal of my life. The big yellow mixing bowl sat in the middle of the table, filled to the brim with well-browned pieces of chicken, stewed until it was almost ready to fall off the bones, whole potatoes, and carrots—with big puffy dumplings, mixed at the bedside, floating on top.

Father had always said grace before meals; always the same twenty-five words, and the ritual was always the same. Mother would look around the table to see that everything was in readiness; then she would nod to Father. That night she nodded to me, and I became a man.

Center Point Publishing
600 Brooks Road • PO Box 1
Thorndike ME 04986-0001 USA

(207) 568-3717

US & Canada:
1 800 929-9108